The Elements of Counseling

Third Edition

D1052327

The Elements of Counseling

Third Edition

Scott T. Meier
State University of New York
at Buffalo

Susan R. Davis
Private Practice

BROOKS/COLE PUBLISHING COMPANY

I(T)P® An International Thomson Publishing Company

Pacific Grove • Albany • Belmont • Bonn • Boston • Cincinnati
Detroit • Johannesburg • London • Madrid • Melbourne • Mexico City
New York • Paris • Singapore • Tokyo • Toronto • Washington

A CLAIREMONT BOOK

Sponsoring Editor: *Eileen Murphy*
Marketing Team: *Jean Thompson & Margaret Parks*
Marketing Representative: *Dawn Beke Burnam*
Editorial Assistant: *Lisa Blanton*
Production Editor: *Mary Vezilich*
Manuscript Editor: *Betty Berenson*
Interior Design: *Wendy LaChance/By Design*
Cover Design: *Cloyce Wall*
Typesetting: *By Design*
Printing and Binding: *Malloy Lithographing, Inc.*

For more information, contact:

BROOKS/COLE PUBLISHING COMPANY
511 Forest Lodge Road
Pacific Grove, CA 93950
USA

International Thomson Publishing Europe
Berkshire House 168–173
High Holborn
London WC1V 7AA
England

Thomas Nelson Australia
102 Dodds Street
South Melbourne, 3205
Victoria, Australia

Nelson Canada
1120 Birchmount R oad
Scarborough, Ontario
Canada M1K 5G4

International Thomson Editores
Seneca 53
Col. Polanco
México D.F., México
C. P. 11560

International Thomson Publishing GmbH
Königswinterer Strasse 418
53227 Bonn
Germany

International Thomson Publishing Asia
221 Henderson Road
#05–10 Henderson Building
Singapore 0315

International Thomson Publishing Japan
Hirakawacho Kyowa Building, 3F
2–2–1 Hirakawacho
Chiyoda-ku, Tokyo 102
Japan

Printed in the United States of America
10   9   8   7   6   5   4   3

Library of Congress Cataloging-in-Publication Data

Meier, Scott T., [date]
    The elements of counseling / Scott T. Meier, Susan R. Davis. -- 3rd ed.
        p.   cm.
    Includes bibliographical references and index.
    ISBN 0–534–34547–6 (paper)
        1. Counseling.   I. Davis, Susan R., [date].   II. Title.
BF637.C6M42   1996
158' .3--dc20   96–23907
                              CIP

In memory of Oliver Meier, 1928–1995

About the Authors

Susan R. Davis is a licensed psychologist who received her PhD in clinical psychology from Southern Illinois University, Carbondale. She is in full-time private practice and adjunct clinical professor in the Department of Counseling and Educational Psychology, SUNY Buffalo.

Scott T. Meier is a licensed psychologist who received his PhD in counseling psychology from Southern Illinois University, Carbondale. He is associate professor and director of training of the Counseling Psychology Program, Department of Counseling and Educational Psychology, SUNY Buffalo.

Preface

Strunk and White's (1979) *The Elements of Style,* a basic introduction to English composition, provided the model for this book. As White notes in his introduction, Strunk originally composed *The Elements of Style* to "cut the vast tangle of English rhetoric down to size and write its rules and principles on the head of a pin." Strunk, a university English instructor, attempted to produce a set of rules to help students avoid basic mistakes in composition. Our purpose is similar: distill the basic elements of counseling and teach what counseling is as well as what it is *not.*

Describing counseling's basic elements is a difficult task. In these preparadigmatic days (Kuhn, 1970; Staats, 1983), counseling lacks a strong consensus about such fundamental issues as the integration of diverse counseling approaches (Ivey, 1980), the usefulness of counseling research to practice (Gelso, 1979, Goldman, 1976), and the specific interaction between various client characteristics and counseling approaches (Krumboltz, 1966). More than 400 types of psychotherapy are said to be operating (Karasu, 1986). This lack of consensus is understandable given the youth of the helping professions and the complexity of the undertaking. Moreover, this disunity is shared by most other branches of psychology and the social sciences (see Meier, 1987). It is important to note that the profession *does* possess a knowledge base; however, theory, research, and practice in counseling often appear fragmented and contradictory.

The issue of disunity is long-standing (see Forsyth & Strong, 1986; Gelso, 1979; Rogers, 1963), but we reiterate the debate because we believe disunity remains a major impediment to progress in the profession. *What* metatheories of counseling eventually develop are probably not so important as *how* they are constructed and presented. Elsewhere (Meier, 1987), we have proposed that an integrated theory should possess four major attributes: (1) the theory should be interconnected, thereby bridging diverse approaches; (2) the theory should be surprising and make strong, unexpected predictions that intrigue counseling researchers; (3) the theory should be readily applicable, specific enough to permit derivation of a technology of counseling; and (4) the theory should be aesthetically pleasing to appeal to current artisans and practitioners of counseling.

If there is so much disunity in counseling, what is there to teach in a basic elements book? Our sense is that much of the disagreement centers on counseling interventions; counselors share greater agreement about the initial stages of counseling. For example, most counselors would agree that information must be gathered about the client and that some degree of rapport must be established between client and counselor. Building on the work of others, this book represents our attempt at delineating these basic elements of counseling practice. Consequently, *The Elements of Counseling* focuses on relationship building and self-exploration, the foundations upon which further intervention is laid.

Although some agreement exists among counselors about initial counseling practices, little consensus has been reached about the timing, sequence, or relative importance of these practices (Goldfried, 1983; see also Cormier & Cormier, 1991). Thus, different counselors and counseling instructors might rearrange substantial portions of this book to suit their preferences and experiences. That makes sense. This book could be read in sections or in sequence, as with a textbook. But this book, like the counseling profession, needs an underlying foundation.

Given the need for brevity and the positive response to the first two editions, we decided to minimize changes in the third edition. We did add a section on client assessment in Chapter 4 and on brief therapy in Chapter 6.

Although suggested by several reviewers, we did not move Chapter 5 ("Counselor, Know Thyself") to the beginning of the book. Counseling theories differ in the emphasis they place on counselor traits and issues and the counselor/client relationship. Counselors who hold counselor characteristics and the counselor/client relationship as central might consider assigning "Counselor, Know Thyself" as the first chapter to be read.

The Elements of Counseling is useful for graduate students in the helping professions (such as psychiatry, psychology, social work, and counseling), as well as for paraprofessionals and peer counselors who are learning basic counseling, communications, and listening skills. Undergraduate students enrolled in introductory psychology or counseling courses will find the book useful in helping them understand the applied aspects of psychology. We also hope experienced counselors and psychotherapists will find these ideas to be valuable reminders.

Although *The Elements of Counseling* presents guidelines and rules, one cannot become a counselor simply by memorizing them. Given the necessarily brief nature of *The Elements of Counseling,* students may overestimate or underestimate their capabilities if they only read this book. Practice in role plays, actual experience, competent supervision, and development of a personal counseling theory are the requisities for competent counselors. Similarly, no claim can be made that this book exhaustively lists all the basics of counseling. *The Elements of Counseling* is designed to function as a brief reference tool. We hope that what the book lacks in comprehensiveness it atones for in simplicity, clarity, and brevity.

The Elements of Counseling has many coauthors. These include past teachers, supervisors, colleagues, and students—the persons who taught us about the elements of counseling. Thank you all. We also acknowledge the helpful comments and suggestions of the book's reviewers (Mildred W. Boyd, Essex Community College; John Garcia, Southwest Texas State University; Joseph Gardner, Middlesex Community College; Lisa Y. Zaidi, Seattle University; Gerald Rabideau, State University of New York—College at Buffalo; Amy L. Reynolds, Fordham University); the production editor, Mary Vezilich, and the editor, Eileen Murphy.

Scott T. Meier
Susan R. Davis
stmeier@acsu.buffalo.edu

Contents

Counselor, Know Thyself   61

A Brief Introduction to Intervention   70

An Approach to Counseling Process

Process produces outcome. The process of counseling—what the counselor and client do in session—influences the outcome, the success of counseling. To master process, beginning counselors must develop a repertoire of helping skills as well as a theory of counseling that directs their application.

How do you go about developing an approach to process? It is no easy task: a recent estimate placed the number of different counseling approaches (such as psychoanalytic, behavioral, and rational/emotive) at more than 400 (Karasu, 1986). Research on psychotherapy outcome shows no broad superiority for any one approach (Smith & Glass, 1977), and no theoretical consensus has yet been negotiated among the major schools.

Needless to say, many contemporary counselors describe themselves as integrative and eclectic. At its best, eclectic counseling involves *doing what works*. Given their training in a wide variety of counseling approaches, eclectic counselors rationally and intuitively select an approach based on the individual needs of their clients. In some cases, this matching process is straightforward. For example, if a client appears at your office with a simple phobia, you could employ systematic desensitization with a strong expectation for success (Wolpe, 1990). However, the match between client and approach is often unclear, and in these instances the weakness of eclecticism is exposed. At its worst, then, eclectic counseling involves guessing or "flying by the seat of your pants."

We describe below twelve approaches to counseling we found to be important, based on a mixture of research and

1

counseling practice, which counselors use to begin and strengthen a helping relationship. Students who master these steps will have acquired a repertoire of basic helping skills.

 ## Make personal contact

The foundation of counseling is the relationship between counselor and client. Different approaches emphasize the counseling relationship to varying degrees, but all practitioners understand that the client and the counselor must first make contact.

Because counselors lack a precise method for describing relationships between people, *making personal contact* is difficult to describe. We have to talk around the topic: making contact means being with the client, touching someone emotionally, communicating. This does not mean the client must immediately develop an intense relationship with the counselor. In fact, the contact may need to be moderate for clients who are afraid of intimacy and personal contact.

Pay particular attention to making contact with your clients during the first session (see Gunzburger, Henggeler, & Watson, 1985; Pope, 1979; Sullivan, 1970). Be open to your clients' lead. If they start to chat, chat for a minute, then return to the business of counseling.

> *Counselor:* Hi, I'm Susan. I'm your counselor.
>
> *Client:* Hello, I'm Bill. It certainly is snowing hard out there.
>
> *Counselor:* Yes, it is. Did you have any trouble getting *(contact!)* here?
>
> *Counselor:* Let's talk about your problems, not the snow. *(too direct)*

Allowing your clients to lead in the initial stages of counseling encourages the development of trust. It also provides information about their agenda and their interpersonal styles (MacKinnon & Michels, 1971).

What could stop you from making personal contact? For example, we begin to feel rushed if we see too many clients consecutively. To avoid this, we set aside 10 to 15 minutes between clients to take notes from the previous session and review notes in preparation for the next hour. Resting between sessions enables us to attend to each client as a unique individual.

 Develop a working alliance

Making personal contact is the first step in developing a working alliance (Greenson, 1965; Zetzel, 1956). The task of the counselor is to engage the client in such a way that both persons are working together to resolve the issues that brought the client to counseling. Such alliances do *not* occur, for example, when the counselor attempts to force the client to change or when the client is unmotivated.

Counselors invite their clients into this working alliance by extending understanding, respect, and warmth. Thus, counselors must be skilled interpersonally.

> *Client:* I'm at the end of my rope. I'm so frustrated!
>
> *Counselor:* This is *really* a difficult time for you.
>
> *Client:* It feels good to hear you say that. None of my friends understand what I'm going through.

Counselors are skilled listeners. By learning about clients through attentive listening and offering acceptance of clients as they are, counselors develop a bond of trust and support. Without this alliance, many clients are unable to change.

 Explain counseling to the client

Researchers call this *role induction* (Hoehn-Saric, et al., 1964; Mayerson, 1984; Orlinsky & Howard, 1978). Clients frequently approach counseling with misconceptions about the process. For example, they may expect counseling to resemble a visit to a medical doctor: diagnosis, prescription, cure. If mistaken expectations are ignored, clients may drop out or fail to make progress.

> *Client:* Doc, why aren't you asking me more questions about my mental illness?
>
> *Counselor:* John, I see counseling being most helpful when you're talking with me about your feelings. That's the best way for both of us to learn about what's going on with you.
>
> *Client:* Well . . . okay.

In this example, the client expected the counselor to lead the counseling session by asking questions. The counselor explained, in essence, that it was the client's job to talk and that it was the current task of counseling to explore feelings.

Beginning counselors should avoid giving explanations until they feel comfortable and knowledgeable enough to do so. Practice first in role plays. *What* to explain depends on such factors as the presenting problem or agency policies; for example, it may be unnecessary to describe the limits of confidentiality regarding potential homicide to a student seeking career counseling. Clients may find it helpful, however, to know that (1) they will do most of the talking, (2) they may experience painful feelings before they begin to feel better, (3) exceptions exist regarding the confidentiality of counseling, (4) persons in counseling are not inherently weak, and (5) most individuals in counseling are quite sane.

Clients may also find it helpful to know that they can take some time to find a resolution to their problems. Many clients approach their first session hoping to find an immediate solution.

> *Client:* I'm glad we could meet today. I'm going home this weekend and my father wants to know what my major is and what kind of job I'm going to get when I graduate.
>
> *Counselor:* Sounds like you're under pressure to make a quick choice.

Client: Sort of—he keeps asking me about it and I just want to be able to tell him something.

Counselor: We might be better off spending more than 50 minutes deciding on your major and your career. It's important to explore your abilities and interests as well as job requirements.

Client: I guess so.

Counselor: It might also make sense to talk about how your father will react to your decisions.

Client: That's true.

In this example, the counselor provided the client with permission to take more time as well as a rationale for why the extra time might be useful.

Role induction may be part of a set of procedures designed to increase clients' expectations for therapeutic gain (Frank, 1971). Clients' (and therapists') belief in the effectiveness of counseling affects its outcome (Rosenthal & Frank, 1958; see also Orlinsky & Howard, 1978, pp. 300–304). Part of the counselor's job, then, is to provide clients with realistic hope for improvement.

Client: I'm getting so afraid of snakes that I won't leave the house much any more.

Counselor: The approach we're going to use—called systematic desensitization—has proven to be highly effective for eliminating phobias like your fear of snakes.

Client: Okay . . . what do we do?

In this example, the counselor described the counseling procedure and its past effectiveness. However, once you become skilled at explaining counseling to your clients, you may be tempted to explain *everything*. This places the counselor in the foreground and encourages clients to hide behind intellectualizations.

> *Client:*   After we argued I went home and kicked the
>   dog. Is that reaction formation or displace-
>   ment?
>
> *Counselor:* Displacement. Instead of expressing your
>   anger to your boss you expressed it to your
>   dog.
>
> *Client:*   Right. I thought it was displacement, but I
>   wasn't sure.

Here the counselor has taken on the position of expert and allowed the client to remove the focus from herself. As an alter-
native, the counselor might inquire how that knowledge (of
classifying "kicking the dog" as reaction formation or displace-
ment) would help the client.

Role induction is one way of organizing or structuring counseling (Shertzer & Stone, 1980). Counselors vary in the amount
of structure they place on the counseling process, but most professionals at least perform an initial interview (to gather information about the presenting problem and background), set goals
(to determine when counseling will be complete), build the
counseling relationship and explore client issues, implement an
intervention, and finally, terminate the relationship.

 Pace and lead the client

Pacing and leading refer to how much direction the counselor
exerts with the client. When pacing a client, a counselor follows
along in terms of the client's expressed content and feeling.

> *Client:*   I put off studying until the last minute again,
> *(sighing)* and then I just hit the books the whole night.
>
> *Counselor:* You crammed again.
>
> *Client:*   Yeah. That's what happened.

Here the counselor paced the client by succinctly restating the
client's concern. Nothing was added to the client's meaning; no

direction was given. Pacing lets the client know that the counselor is listening and understanding.

Reflection of feeling and *restatement of content* (Egan, 1990) are two effective methods of pacing clients. Reflection of feeling refers to the counselor's recognition of the client's feeling and a subsequent mirroring of that feeling. With restatement of content, the counselor notices the client's thoughts and restates that content. Reflection of feeling and restatement of content build rapport between counselor and client by developing a consensus about what the client is thinking and feeling. For many people, having another person listen deeply makes those experiences more real. In fact, some clients seem unaware of what they are experiencing until they hear a restatement or paraphrase.

> *Client:*   I was so mad I just got up and left.
>
> *Counselor:*   You were really angry with your husband.
>
> *Client:*   Yes . . . I guess I was. I just realized how angry I was—and still am!

Pacing and leading are other ways of discussing *timing*. Through experience and through watching the individual reactions of clients, counselors develop a sense for when they should direct the counseling process.

> *Client:*   I'm stuck. I just can't keep putting off studying like this.
>
> *Counselor:*   I wonder if your cramming has anything to do with your desire to avoid responsibility for failing the course.
>
> *Client:*   Well . . . my dad thinks I get Cs because I party like he did in college, not because I don't have the ability. Hmmm . . .
>
> *Counselor:*   (silence)

In this example, the counselor intervened at a point when she believed the client was ready to hear a new message. Once the client made the connection, the counselor backed off and allowed the client to think about the new material.

Psychologically sophisticated clients progress more quickly because of their greater familiarity with the counseling process. Highly motivated clients can better tolerate the pain and ambiguity often present during the course of counseling. Strong rapport also signals the counselor to move ahead on client issues.

Counselor: We've spent several sessions speaking about how you can adjust to your illness . . . but we haven't spoken about death.

Client: I know.

Counselor: *(silence)*

Client: I know . . . I know I have to start talking about dying.

The counselor recognized this individual's self-awareness of his fear of death and gently brought it to his attention. The client's awareness of the importance of talking about that fear helped him to begin the process.

Pacing occurs most naturally when client and counselor are similar on such variables as socioeconomic status, personal values, cultural background, and life experiences. Similarity encourages understanding based on intuitive insight into the other's experience. The effectiveness of peer counseling (Varenhorst, 1984), group counseling (Yalom, 1995), and paraprofessional counselors (Christensen & Jacobson, 1994) may be partially due to the pacing that naturally occurs in such homogeneous settings. The status of professionally trained counselors may increase credibility, but at a cost to their capacity to match and pace clients.

An important contraindication for leading is your state of stress: stressed counselors may be too inclined to take risks and push their clients (Freudenberger, 1974; Meier, 1983; for an introduction to the area of impaired professionals, see Kilburg, Nathan, & Thoreson, 1986). In general, when counselors lead too much they lose their clients, figuratively and literally.

Introducing three or four major insights during one session may seem like brilliant counselor strategy, but clients are likely to remember only one or two themes. Leading clients into a sensitive area too soon can be risky. One beginning counselor felt he had zeroed in on his client's major issues and began to share his interpretations during the first session. He assumed the client would recognize the truth of these insights and change. Instead, the client became frightened and dropped out of counseling.

 Speak briefly

In general, counselors should talk less than their clients. Except when summarizing, communicate in one or two sentences. Unfortunately, even the best counselors can get carried away.

> *Client:* My dad got angry at me, so I went to the store and got something to eat.

> *Counselor:* So your father lost his temper again, and as usual, you got a stomachache when you started to feel anxious about your dad's yelling. I wonder if you were eating as a way to manage your anxiety or whether you just needed to get out of the house? Is there something else you can do to change that the next time your dad becomes angry?

In this example, the counselor shared all of her interpretations and solutions at once. If you were the client, how would you feel under this bombardment of questions?

Beginning counselors typically have difficulty being brief with talkative clients. After a goal or theme has been established (for example, to discuss the client's relationship with his father), it's okay to bring a wandering client back to the main issue.

> *Client:* So when my dad yelled at me for the third time that day, I really started to feel anxious. I went down to the corner market and bought something to eat so my stomach wouldn't hurt

so much. I didn't want to study, so I called my boyfriend from the store phone and we got into an argument. I haven't talked to him since then but I'm wondering if—

Counselor: Could we stop a moment and talk about your feelings about your father?

One technique counselors employ to show they are listening and yet stay out of the client's way is *minimal encouragers* (see Egan, 1990). Minimal encouragers are phrases like "uh-huh" and nonverbal gestures like head nods.

Client: I know I have to start talking about my illness.

Counselor: Yes.

Client: And it's hard to think straight with the pain. It hurts my family to see me in pain.

Counselor: *(nodding)*

Client: They want so badly to help me that we haven't even begun to say good-bye.

This counselor's use of minimal encouragers appears effortless. Nevertheless, the client may be benefiting a great deal from exploring personal issues without counselor interruption.

6   When you don't know what to say, say nothing

Particularly in counseling, silence is golden. Perhaps because silence can indicate awkwardness in social conversations, beginning counselors often feel uncomfortable with silence and quickly fill the gaps between client statements.

Client: So I got laid off after 15 years of service.

Counselor: Fifteen years of hard work . . .

Client: Yeah . . . *(silence)*

> *Counselor:* And you're pretty angry at the company too, I imagine.

The client might have been feeling angry but could also have been experiencing sadness about the job loss or happiness about job accomplishments. However, the counselor's premature statement effectively stopped the client's internal processing of thoughts and feelings. Learn to notice when people are absorbed and dealing with their issues—a sign of real work in counseling.

Apart from the initial interview, it is largely the client's job to talk, not the counselor's. The counselor may sit quietly during silent periods and wait for the client to resume speaking; silence here is not withdrawal. The best therapists, like good referees in sporting events, work in the background. Referees and therapists tend to be most noticed when they are making mistakes.

> *Client:* After I failed that exam I was so angry that—
>
> *Counselor:* You wanted to drop out of school again.
>
> *Client:* —uh . . . no, I wanted to try twice as hard next time.

Not only did the counselor in this example interrupt, but he was wrong; and the former is more serious. Counselors think of themselves as helpers. Listening helps, too—sometimes it's all the help clients need.

7 You may confront as much as you've supported

Confrontation in counseling does not mean opposing the client but pointing out discrepancies between clients' goals and their actions (Shertzer & Stone, 1980). In a way, confronting the client is a way of saying, "Stop a minute! Look at what you're doing."

Beginning counselors typically find confrontation difficult. A good rule of thumb is that you can confront as much as you've supported.

> *Client:*   I just don't have the willpower to eat more.
>
> *Counselor:*   I'm puzzled by that, Joe. When you started counseling two months ago you said that you *really* wanted to get back to your normal weight. That sounded like "willpower" to me.
>
> *Client:*   Well . . . I do want to gain weight. I just can't do it.
>
> *Counselor:*   What's stopping you?

Support and empathy are the foundation upon which the counseling relationship is built. Consequently, confrontation is unwise during the early stages of counseling. However, once you establish a bond, confrontation may increase client self-awareness and motivation to change (Egan, 1990).

 ## 8   If you want to change something, process it

At this chapter's beginning, we described *process* as what occurs during the counseling session. Counselors also employ the word *process* to describe the act of *talking about* something that is happening in the session. Thus, processing refers to discussion of client and counselor feelings about an event or an aspect of the counseling relationship.

Counselors contrast *process* with *content* (Pope, 1979), the latter referring to the overt topic of the conversation.

> *Client:*   The problems in my marriage are all my wife's fault.
>
> *Counselor:*   She has . . .
>
> *Client:*   Started a job. She's not spending any time with
> *(content)*   me or the kids anymore.

The content in the preceding example has to do with the client's belief that his wife is the cause of his problems. In contrast, a process comment often focuses on immediacy—that is, what the client and counselor are *feeling and experiencing at that moment.*

> *Client:* I don't know what else to say.
>
> *Counselor:* Seems that it's difficult for you to talk with me.
> *(process)*
>
> *Client:* Yeah. My wife told me I should see you.
>
> *Counselor:* So you're not here because you want to be.
> *(process)*
>
> *Client:* Yeah . . . we've been fighting a lot. I'm feeling
> *(sighing)* kind of shaky about what's going to happen.
>
> *Counselor:* Tell me more about that "shaky" feeling.
>
> *Client:* I don't know. I love her, but I'm really upset about this work thing.
>
> *Counselor:* Are you feeling shaky right now?
> *(process)*
>
> *Client:* Yes . . . I don't know what else to do to make it work.

In this example, the counselor pointed out the salient aspect of the process: the client wasn't talking. By openly acknowledging the client's discomfort, the counselor helped the client to relax and open up.

9   Individualize your counseling

Each counselor eventually develops a personal style of counseling. This style may be based on a particular theoretical approach or on experience gained in a particular work setting. Remember, however, that you must adapt those general rules and personal techniques to each client.

Explaining exactly *how* to individualize your counseling approach is very difficult because little consensus exists among counselors on this subject. Counselors tend to build personal frameworks for gauging how best to modify their counseling approach. Some counselors assess the psychological sophistication and level of motivation of their clients; others observe social maturity and intelligence. If the client has had previous counseling, you may be able to tailor your approach on the basis of what worked and what didn't.

Counselors observe clients' use of language in an effort to conduct sessions at a matching conceptual level. Be especially careful to avoid jargon that would confuse clients.

> *Client:* I've been really down lately—I'm getting less pay because my hours at work have been cut during winter. I've had to stop jogging, too, because of all the snow we've had.
>
> *Counselor:* You're down in the dumps.
>
> *Client:* Yeah . . . it's depressing.

Not

> *Counselor:* I suspect your decreased rate of receiving positive reinforcement is directly related to your depressive feelings.

Individualize you must. You may alter your language, your posture, even which counseling approach you employ. By doing so you will create a unique application of your counseling style for each client. Remember, you must talk with your clients, not to them.

 ## Notice resistance

Resistance refers to an obstacle—presented by the client—that influences the process of counseling. Resistance may be a key indicator of the client's readiness for change and the types of interventions the counselor should employ.

Counseling theorists construe resistance differently, and most acknowledge the inevitable presence of resistance. Psychodynamic counselors, for example, view resistance as an attempt to keep anxiety-provoking material from awareness; social learning counselors see resistance as stemming from fear of the consequences of changed behavior (Hansen, Rossberg, & Cramer, 1994; Shertzer & Stone, 1980). Mahoney (1987) suggests resistance to change "is a natural expression of self-protection when core ordering processes are challenged" (p. 15); in this context, resistance is something to be respected, understood, and, when appropriate, explored.

> *Counselor:* Gloria, last week we agreed you would speak with your mother about some of the issues we discussed. But you haven't mentioned anything so far today about that.
>
> *Client:* Oh . . . did I say I'd speak with her? I guess I forgot.

Examples of resistance include an abrupt change of topic and forgetting important material. If the client continues to make progress despite such behavior, simply make a mental note of the resistance and go on. However, if you encounter repeated resistance, process it with the client at a time that seems right to you and at an emotional intensity that fits the client. This discussion may be all that's necessary for the client to move forward.

> *Client:* Oh . . . did I say I'd speak with her? I guess I forgot.
>
> *Counselor:* That's strange—I know you really wanted to talk with her at the end of last week's session.
>
> *Client:* I guess I did.
>
> *Counselor:* How do you feel as we talk about it now?
>
> *Client:* Well . . . scared, really. I realized how angry she'll be if I confront her about abusing me while I was a kid.
>
> *Counselor:* You're frightened about confronting her.
>
> *Client:* Yes . . . but I want to do it. That seems like the only way for me to resolve it.

With *extreme resistance,* the counselor returns to pacing the client by decreasing the emotional intensity of the session or by changing the topic (Brammer & Shostrom, 1989). In essence, the counselor gives the client permission to deal with the intense material when the client is more able to do so.

 ## 11   When in doubt, focus on feelings

No matter what their theoretical orientation, counselors often focus on the feelings of clients (Anton, Dunbar, & Friedman, 1976; Pope, 1979). Counselors trust clients' feelings— particularly as expressed on the nonverbal level—as indicators of salient issues.

> *Counselor:*   I have a sense that it's really important for people to like you.
>
> *Client:*   Yes—isn't it for everyone?
> *(tensing)*

This client responded defensively, interpreting the counselor's statement to mean that such a need was abnormal or inappropriate. The counselor might follow up by asking the client what he feels when he wants people to like him.

Learning how to recognize and express feelings challenges many clients (Cormican, 1978). Clients may begin counseling unable to recognize their feelings or to describe them in more than a cursory manner.

> *Counselor:*   How do you feel about losing your job?
>
> *Client:*   Bad. I'm very upset.

Although this client can label the intensity and direction of the feeling, descriptions like "bad/good" provide very little detail. When clients use nondescript words like *bad, good,* or *upset,* ask them to elaborate. The client in the preceding example might be feeling angry because she has to go through the job-hunting process again or sad because she believes that the job loss indicates failure on her part.

Counselors typically search for the "Big Four" of feeling words: (1) *anger,* (2) *sadness,* (3) *fear,* and (4) *joy.* Beginning counselors are often taught to help clients recognize these feelings and the reasons for them; for example, I *feel* angry *because* I didn't get the job (Egan, 1990). An ability to recognize these feelings in clients (and to help clients become aware of them) is a sign of progress in the beginning counselor.

Counselors also attend to feelings because clients seek counseling primarily to alleviate psychological pain. Helping clients pay attention to their feelings can increase their motivation to change. Fully experiencing feelings may bring insight and relief as well.

Experienced counselors can determine clients' feelings by paying attention to how *they* feel as clients talk. By sharing their reactions to clients' situations, counselors can increase clients' experience of their feelings in the here-and-now.

> *Client:* Even though it's been a year since my father died, I still feel a hole in my life.
>
> *Counselor:* As you spoke about your father I felt sad.

This technique may help a client who is having difficulty experiencing or recognizing feelings. The counselor models the feeling and reassures the client that such a feeling is okay.

A few clients see expression of feelings as an all-or-nothing proposition. These clients model the volcano theory of emotion: they express no affect until their feelings build up and explode.

> *Client:* Jim, my coworker, just kept teasing and teasing me until I couldn't stand it anymore and I screamed at him.
>
> *Counselor:* You were feeling very angry. How did he react when you yelled at him?
>
> *Client:* He was surprised.

Assertiveness training, where clients learn how to express feelings to others, may be appropriate for such clients (Eisler,

1976). Clients may also learn to express emotions at a moderated level by practicing such expression in counseling.

— Broly goals into manageable parts.

 12   Plan for termination at the beginning of counseling   *— Create expectation.*

Termination refers to the process that occurs at the end of counseling. At the beginning of counseling, client and counselor should reach at least a tentative understanding about when and how counseling will finish.

Counselor: When will we know counseling is complete?

Wife: When we're not fighting anymore.

Husband: At least not as often as we are now, which is daily.

Counselor: If you fight only once a week, then counseling has worked?

Both: Yes.

Planning for termination means that explicit goals have been set (see Cormier & Cormier, 1991, for specific information about goal setting). Goals should become clearer and may be revised (along with the termination plan) as clients move deeper into self-exploration.

Client: I don't want to be just emotionally independent of my family—I want to have my own apartment too.

Counselor: Moving out would signify a real step toward independence for you.

Client: Yeah. If I did that, I'd be satisfied.

Termination should be planned, not abrupt; your comfort with termination will influence this process. Remember that it is important to say good-bye, consolidate the counseling experience, discover what counseling meant to the client, and discuss future situations.

> *Client:* So it looks like I'm going to go to college after all.
>
> *Counselor:* I'm delighted. You worked very hard to sort things out with yourself and your family.
>
> *Client:* I certainly appreciate your help and listening to me while I was so confused about what to do.
>
> *Counselor:* Well, I enjoyed listening to you. I'm glad you're ready to go, but sad we won't be meeting anymore.
>
> *Client:* Me too . . . but I know I can talk things out with my parents in the future.

Clients say good-bye in different ways. Some persons cling to the relationship and try to avoid ending; some deny any feelings; some cancel the last session. Talk with clients about how they feel about terminating. All this can be made simpler by establishing, at the onset of counseling, the conditions under which counseling will end.

Counselors often schedule several sessions to terminate. With some clients, they may schedule a monthly meeting to maintain progress.

Strategies to Assist Clients in Self-Exploration

This chapter provides additional guidelines about a particularly important aspect of the counseling process: client self-exploration. Self-exploration refers to the elaboration and deepening of self-awareness and self-concept that occur as clients speak about themselves. Self-exploration provides information about what must be done for change to occur, and it can be therapeutic in and of itself. Once clients understand which behaviors will produce the desired outcomes, *some* clients change relatively quickly (see Butcher & Koss, 1978; Crits-Christoph, 1992).

Facilitating client self-exploration depends on what you *don't do* as well as what you *do*. Beginning counselors often work too hard: they feel responsible for *helping* clients, and they take that task as a mandate to give advice, ask scores of questions, and solve problems. In contrast, experienced counselors initially stay in the background and help clients thoroughly understand themselves and their problems.

13 Avoid advice

Friends and family members give advice; counselors generally don't, particularly in the initial stages of relationship building. However, many counselors (beginning and advanced) appear to believe that they must offer counsel.

Counselor: What brings you here today?

> *Client:* Basically, I'm having trouble with my boyfriend.

> *Counselor:* Have you tried talking with him?

Even though the counselor's response is a question, it implies advice (go talk with your boyfriend). Advice is an intervention and as such should be avoided until counselor and client establish a trusting relationship. However, many counselors completely avoid advice because (1) clients have already tried these simple strategies, and (2) the advice has already failed or the client wouldn't be in counseling.

> *Client:* I'm having trouble with my boyfriend.

> *Counselor:* Tell me about the trouble you are having.

It's a much better strategy to encourage self-exploration, as in the preceding example, than to engage in premature problem solving. Because friends and family frequently provide advice to clients, your advice is likely to be old news. Also, some clients resist advice because it is externally generated; these individuals expect and want to solve their own problems.

Don't confuse giving information with giving advice (Cormier & Cormier, 1991). When you give advice, you typically provide clients with specific actions to perform. Information, on the other hand, consists of knowledge, alternatives, or facts that clients may find useful in their decision making. As explained in the following section, the usefulness of providing information also depends on when it is done.

 ## 14   Avoid premature problem solving

Problem solving too early in counseling usually fails (see Egan, 1990). By *problem solving,* we mean efforts by the counselor to create alternatives and suggest strategies to resolve clients' problems. Clients may resist promising solutions or already know what to do; even if clients implement a counselor's suggestion, they have only the counselor to blame or praise. Clients bear the ultimate responsibility for change. Similarly, the problems clients are facing are unlikely to be their last, and a solution selected by the counselor will not help clients learn how to handle future difficulties.

Client: I just can't find the willpower to start studying a week before an exam.

Counselor: Maybe if you drew up a schedule with regular study times you could study more consistently.

Client: That would work, but I doubt that I could stick to the schedule.

Often clients know how they could obtain desired goals and outcomes but have no confidence in their personal capacity to act in the required manner. These clients lack what Bandura (1977) calls self-efficacy, a belief that one can perform a behavior that leads to a desired outcome. For example, alcoholic clients who have low self-efficacy for stopping drinking may relapse when they are repeatedly offered drinks at a party. Self-efficacy expectations influence whether clients initiate certain behaviors and persist in the face of obstacles.

Rather than problem solve early in counseling, counselors do better to help clients define problems fully. For example, you might explore what solutions have already been tried. This may help counselor and client to better understand why previous efforts have failed or have exacerbated the original problem.

Counselor: You've said that you plan to study a week before the exam, but you don't follow through. What stops you?

Client: I start to study—I stay up all night and read, but after a couple of nights of that I just don't feel like working anymore.

Counselor: Sounds like you just get worn out.

Client: That's right.

A client's solution to a problem can sometimes produce additional problems or exacerbate the original difficulty (Watzlawick,

Weakland, & Fisch, 1974). In this example, the client became physically fatigued because of poor study habits and lost motivation to study further.

 Avoid relying on questions

Faced with their first real or role-play clients, new counselors may simply ask questions. Beginners use such questions to elicit more information or to suggest advice. It is certainly permissible to ask about specific information or to seek elaboration. However, clients may perceive a sequence of questions as threatening.

> *Client:* So I just left after we started to argue.
>
> *Counselor:* You were angry?
>
> *Client:* Yes.
>
> *Counselor:* Why didn't you stay and work things out?
>
> *Client:* I was too angry to talk.
>
> *Counselor:* You felt out of control?

Questions, particularly "why" questions, put clients on the defensive and ask them to explain their behavior. Questions keep the counselor in sole control of counseling and may inappropriately lead clients. Statements that *reflect* the content or feeling clients have expressed, however, do not imply direction by the counselor.

> *Client:* I just left after we started to argue.
>
> *Counselor:* You got up and left the room.

Counselors also differentiate between open and closed questions (see Egan, 1990). Open questions seek elaboration by clients; closed questions ask for specific information and may be answered in a word or two. Counselors particularly should avoid sequences of closed questions when they are facilitating client self-exploration.

> *Client:* I just left after he started to swear at me.

> *Counselor:* How did you feel then?
> (*open*)

> *Client:* I was really angry. He had no right to speak to me that way—we had been close friends for years.

> *Counselor:* Did you swear back?
> (*closed*)

> *Client:* No.

The open-ended question in the preceding example encouraged the client to describe her feelings and the reason she was angry. The closed question provided no pertinent information. Unless you need very specific information, stay with open-ended inquiries.

 ## 16   Listen closely to what clients say

As a counselor, you will listen for the meaning of the words your clients use. You can, however, also listen for individual words and grammar. Words and the manner in which they're employed can provide important clues about how clients view the world (Bandler & Grinder, 1975).

> *Client:* My children always disobey me.

> *Counselor: Always?*

> *Client:* Well . . . not always. They usually get up and get dressed for school without any trouble.

A word like *always* indicates a distortion of reality, a predisposition to perceive different events in the same way. In the preceding example the client didn't really mean "always" when he described his children's misbehavior. His recognition of this distortion of reality may help him perceive his children differently or discover a new solution.

Similarly, counselors should determine what clients mean when they use words like *must* and *should*.

Client: I *must* get an A on this test!

Counselor: There's a feeling of desperation in your voice.

Client: Yes. Without this grade I'm doomed.

Ellis is particularly vigilant in listening for words like *must* and *should* because he views them as signals of irrational beliefs (Ellis & Grieger, 1977; Ellis & Harper, 1976). Ellis works to detect thinking that has no basis in reality and that tends to catastrophize events.

When clients speak, they may fail to describe fully their experience (Bandler & Grinder, 1975). What is deleted or missing may be indicative of the client's problem.

Counselor: Without an A you're doomed to what?

Client: Well . . . I don't know. I won't maintain my 3.5 grade point average.

Counselor: So . . . .

Client: I wouldn't get into med school.

Counselor: And then?

Client: Well . . . I'd be very disappointed. I guess I'd have to find another career.

Asking clients to present a fuller description of themselves is a method of facilitating self-exploration *and* helping clients change.

17 Pay attention to nonverbals

People communicate with each other by paying attention to the verbal content of messages and to the most overt nonverbal messages (for example, smiling, frowning, making a fist). But the more subtle nonverbal components of communication—

tone of voice, facial expressions, eye contact, and body mo-
tion—convey equally rich information. The nature of nonverbal
communication can be influenced by culture (Sue, 1990a), but
unfortunately there is no Rosetta stone that can be used to de-
cipher the meaning of nonverbals with any particular client. No
one can tell you exactly what to look for, but the importance
of observing others' nonverbals, in role plays and everyday situations, cannot be overemphasized.

One particularly effective way that counselors confront clients
is to point out discrepancies between nonverbal and verbal communication.

> *Client:* I was really upset when she said she didn't like
> *(smiling)* me.

> *Counselor:* So you were upset . . . but I notice you were
> smiling when you said you were upset.

> *Client:* I was? Well . . . it is kind of hard for me to ad-
> mit she upset me.

When faced with conflicting messages on nonverbal and verbal levels, counselors tend to trust nonverbal communication as
more indicative of basic feelings. The assumption is that it's easier to censor verbal than nonverbal communication. Counselors
often help clients become aware of the nonverbal level.

> *Client:* I was really upset when she said she didn't like
> *(smiling)* me.

> *Counselor:* You were upset . . . but I notice you were smil-
> ing when you said you were upset.

> *Client:* I was? Well . . . it is kind of hard for me to ad-
> mit she upset me.

> *Counselor:* How do you feel when you smile like that?

> *Client:* I don't know. . .

> *Counselor:* Smile again, just like you were.

> *Client:* Okay . . .
> *(smiling)*

> *Counselor:* How do you feel now?
>
> *Client:* Well, maybe a little safer . . . maybe you'll still like me even though she upset me.

This client was able to learn from exploration of her nonverbal movements. She learned that she smiled to hide her embarrassment about her feelings.

Counselors also express different messages on verbal and nonverbal levels. For example, beginning counselors sometimes experience difficulty being congruent in their communication (Shertzer & Stone, 1980).

> *Counselor:* Your dad died 2 years ago today. *(smiling)*
>
> *Client:* Yes . . . but I still feel sad about it. *(confused)*

Some counselors simply feel more comfortable if they smile while around others. Such habits, however, can send mixed signals to clients. Videotaping and audiotaping sessions can help you discover any idiosyncratic nonverbal expressions that interfere with the counseling process.

 ## 18   Focus on the client

Clients often talk about other people. Others may be seen as sources of one's troubles or as standards for behavior.

> *Client:* My mom thinks I should go into social work.
>
> *Counselor:* How do you feel about that?

With this client, the counselor has the choice of following the other person's feelings ("What makes your mother think you should be a social worker?") or of following the client's feelings ("How do you feel about your mother's belief?"). Almost invariably, counselors return the focus to the client. The client is the object of change in counseling, not other people. On the other hand, when others may influence the process of change, as in family or marriage counseling, counselors arrange for those relevant individuals to be present.

Many clients experience difficulty in talking about themselves. Counselors can help these clients through appropriate self-disclosures (Jourard, 1971) or self-involving responses (McCarthy, 1982; McCarthy & Betz, 1978). With self-disclosure, the counselor reveals a personal feeling or experience of his or her own; with self-involving responses, the counselor responds personally to a statement by the client. In both cases, the intent is to help clients get in touch with *their* feelings.

> *Client:* So I got laid off after 15 years on the job. Now I've got to begin job hunting and I haven't the slightest idea where to start.

> *Counselor:* I know what you mean. I lost my last job when
> *(self-* the agency had its funding cut. I was
> *disclosing)* depressed.

Or

> *Counselor:* I feel sad that you lost your job.
> *(self-*
> *involving)*

> *Client:* Well . . . I have been really down. I've been trying to think positively and ignore those feelings because I've got to get a job as soon as I can.

Self-disclosing and self-involving statements encourage clients to reciprocate. Self-involving responses in particular may help clients become aware of their feelings and speak more about themselves. However, counselors' self-disclosing and self-involving statements should be employed sparingly (to maintain the focus on clients) and at a matching level of intensity. As an example of the latter, it would be inappropriate for a counselor to disclose her or his current feelings of depression to a person seeking career counseling. A more appropriate self-disclosure might involve the counselor's revelation of past feelings of confusion and anxiety surrounding career decisions.

Be concrete

Some of counseling's working materials, such as feelings and thoughts, are intangible. For example, counselors have compared feelings to the wind, suggesting that clients should notice emotions and allow them to pass through and change. Skilled counselors help clients make their feelings and thoughts concrete.

> *Client:* Even though it's been a year since Dad died, I still feel a hole in my life.
>
> *Counselor:* Tell me about that hole.
>
> *Client:* It's kind of an emptiness . . . a numbness.
>
> *Counselor:* You're holding your stomach as you talk about the emptiness. Hold your stomach a minute . . . tell me what you feel.
>
> *Client:* I feel . . . sad.

Grounding feelings within the body is one method of making feelings concrete. As suggested in this example, the client might learn to pay attention to his stomach whenever he feels sad. Gestalt counselors use clients' body movements, such as repeatedly making a fist or striking a pillow, to help strengthen clients' experience of feelings.

Concreteness becomes especially important when discussing clients' behavior and goals. What exactly do clients want to change? How will client and counselor know when counseling is finished?

> *Client:* I want to do better in school.
>
> *Counselor:* What do you mean by "better"?
>
> *Client:* I don't know . . . maybe get a 3.0 average.
>
> *Counselor:* You want a 3.0. Are you speaking about just this quarter or the whole year?
>
> *Client:* Just this quarter.

Clients sometimes have difficulty going beyond broad ideas and intellectualizations. Counselors help these clients by asking about specific events.

> *Client:* You know, if my husband would just complete the separation process with his family, we wouldn't argue so much.
>
> *Counselor:* Give me an example.
>
> *Client:* Well, last week we started to argue about the phone bill, and immediately he calls his mother for advice. Then his parents are angry at me, too, and I really feel anxious.

In this example, the counselor learned that the family of the client's spouse becomes inappropriately involved in their marriage. From such information, the counselor might decide to pursue marriage counseling or help the client plan for such events in the future.

 ## Listen for metaphors

A metaphor is a figure of speech, containing an implied comparison—expressing an idea in terms of something else. Clients will occasionally offer metaphors about their issues. Listen for and accept these gifts—they are worth their weight in gold.

> *Client:* I'm making progress in counseling, but it sure is hard.
>
> *Counselor:* You're making a lot of effort.
>
> *Client:* Yeah . . . it reminds me of a 15-round boxing match, and we're only in round 3.

Metaphors are easy to remember. By using the boxing analogy, the counselor could ascertain the client's perception of progress in counseling (for example, round 7 would indicate further progress than round 3). Another client compared her problems to a box: the size of the box indicated the client's perception of the current intensity of her problem. When the client felt she had to place her attention elsewhere (for example, to study for a test), she "put the box in the closet for now."

 21   Summarize

Summarizing refers to a brief review of the major issues in counseling. You might summarize important issues during the counseling session, at the end of a session, or at the final counseling session. Client or counselor may summarize. Client summaries present counselors with an opportunity to check their ideas about the counseling process with clients.

> *Counselor:* So if you were to sum up what has gone on in counseling so far, how would you describe it?
>
> *Client:* Well, I certainly understand how my parents and I get started on arguments. Everybody gets so involved in presenting their case no one listens to the other. I know I have to point that out to them, and notice it myself, or I'll get really frustrated in a hurry.
>
> *Counselor:* That certainly fits with what I've heard you say in here.

Summarizing is another method of structuring counseling. Summaries often describe important themes, keep track of change in counseling, and help connect related issues.

> *Client:* Any time I start a new class I feel really anxious. I just don't have any confidence in myself about taking risks.
>
> *Counselor:* I wonder if your lack of confidence has anything to do with your mother's overprotectiveness. You've talked about how she rarely let

you try new experiences on your own as a
child. I wonder if the two are related . . .

Client: That makes sense.

By summarizing a theme (the client's mother protected her
from taking risks throughout childhood) that had appeared several times previously, the counselor helped the client gain insight
into her current feelings of anxiety.

A Few Mistaken Assumptions

Counseling remains a poorly understood profession among laypersons. Many people still see counseling, for example, as a place only for the sick, weak, or severely disturbed. Beginning counselors must *unlearn* some of the common assumptions they bring with them to training. In a sense, this book is a role induction for beginning counselors and a role reminder for experienced therapists.

 ## Positive thinking does not equal rational thinking

Some beginning counselors mistakenly equate Ellis's concepts of rational and irrational thinking with positive and negative thinking. Positive and negative thinking usually refers to beliefs that one will encounter good or bad fortune. Irrational thinking refers to a belief, unsupported by objective evidence, which leads to a painful feeling (Ellis & Grieger, 1977).

> *Client:* I *must* get an A on all of my final exams or I'm doomed.
>
> *Counselor:* You're saying you must be perfect in all of your classes.

As we noted in the previous chapter, words like *must* indicate irrational beliefs. In this case, the counselor interpreted the client's obsession with perfect grades as an irrational belief.

Although counselors may help clients improve their self-esteem and self-confidence, they should avoid trying to talk clients into improvement. Particularly with persons whose problems are long-standing, verbal persuasion, by itself, is one of the weakest methods of promoting change (Bandura, 1977).

> *Client:* For years I've been afraid people will laugh at me if I speak in class.

> *Counselor:* You just have to believe people are going to
> *(ill-advised)* like you.

> *Client:* I know that, but I can't do it.

Instead, counselors challenge clients to provide proof that their catastrophic beliefs will come true. Frequently counselors work with clients to design homework to test the rationality of their beliefs.

> *Client:* I'm afraid people will laugh at me if I speak in class.

> *Counselor:* What could you do to test that belief?

> *Client:* Well . . . I could ask a question.

> *Counselor:* That makes sense to me. How about asking one question in each of your classes and seeing how people respond?

At the following session, counselor and client would discuss the results of the homework. In this example, the client would likely learn that no one laughed at her questions. Even if a classmate later ridiculed the client, the counselor might acknowledge the unpleasantness of that event but point out that the client did survive it without any real catastrophic outcome.

 ## 23   Agreement does not equal empathy

Some beginning counselors interpret *empathy* to mean agreement or sympathy. Empathy refers to the comprehension of the subjective world of clients (see Egan, 1990). Agreement suggests that the counselor approves of the client's behavior, and sympathy indicates that the counselor feels sorry for the client.

Client: I was really angry when he said to go home.

Counselor: You felt angry with him.
(empathy)

Counselor: Good for you. I'm glad you left!
(agreement)

Counselor: He is awfully mean to you!
(sympathy)

Friends and family provide agreement and sympathy. Counselors provide empathy to help clients explore their problems and become aware of their feelings and thoughts. In that way, clients begin to understand what they need to do to change.

 ## Do not assume that change is simple

Clients typically expect to change some aspect of their lives so they will feel better. However, matters are seldom so simple. Human behavior has multiple causes, and no counselor can always be aware of all the factors helping and hindering change. Such factors as significant others, biological influences, and individual differences in response to therapy can all contribute to the complexity of the counseling process.

Client: I started on the homework, but Mother says she doesn't think it will work. She says I should stop counseling.

Counselor: Well . . . that's the first I've heard about your
(surprised) mother since our first session.

Client: I think she might be right.

The counselor had been proceeding on the assumption that changes in the client's behavior would resolve the presenting problem. However, the client's mother has intervened and may now need to be included in the counseling process.

Counselors sometimes do marriage and family therapy because they view individuals' problems as relating to the social systems to which they belong (deShazer, 1982; Minuchin & Fishman, 1981). Change is not simply a matter of altering the "misbehaving" person. An adolescent may act out at school to *stabilize* her family system: her feuding parents, considering divorce, might table their disagreements to handle the current crisis. More effective counseling would focus on the entire family rather than just on the client.

25 Make psychological assessments, not moral judgments

One of the most difficult tendencies to alter, in some beginners, is judging people. Being judgmental involves moral or ethical assessments.

> *Client:* So I began to drink again after I lost my job.

> *Counselor:* You really let your family down there, didn't you?

Instead of judging whether behavior is right or wrong, counselors should assess clients in terms of psychological theory and practice. Such assessments might involve inquiry into family background, educational and employment experience, psychopathology, intellectual abilities, physical health status, and situational influences.

> *Client:* So I began to drink again after I lost my job. I sat at home a lot, and I just kept getting more irritable.

> *Counselor:* It sounds as if you felt very stressed and then started to drink again.

In this example, the counselor suggests a psychological cause for the client's drinking, a cause that the client can influence and for which the client bears responsibility. No condemnation of the person is involved.

Sin and forgiveness are religious terms, not counseling concepts. Many clergy perform pastoral counseling in which they serve not as preachers but as skilled listeners and helpers. Some beginning counselors, however, view counseling as a means of promoting religious values to solve clients' problems. Although it has well-intentioned and skilled followers, religious counseling is a contradiction in terms if practitioners intend, through their counseling, to save or convert their clients. Clients have a right to their personal values.

 ### 26 Do not assume that you know clients' feelings, thoughts, and behaviors

When we talk with other people, ordinarily we assume that we know what they are feeling or thinking. Experience in counseling demonstrates the fallacy of this assumption.

>   *Client:* So I just left after we started to argue.

>   *Counselor:* You were very angry.

>   *Client:* No, I had to leave for work in ten minutes and I knew we'd never resolve anything in that amount of time.

In this example, the client corrected the counselor's misperception. In practice, many clients will tell you when you are wrong. Listen for these corrections.

To avoid the feeling that they know everything, counselors proceed with a sense of tentativeness (Egan, 1990). They act as though they might have misunderstood, and they ask their clients for confirmation of what they know. They verify their assumptions about clients' feelings, thoughts, and behaviors.

>   *Counselor:* You seem sad as you speak of your father. Is that how you feel?

>   *Client:* No, I didn't feel sad as much as I felt angry about getting into another argument.

Counselors express their tentativeness in this form: a reflection of feeling or content followed by a question. Even if the counselor's reflection misses, clients will respond with further elaboration.

Tentativeness reminds counselors of their ignorance of clients' subjective worlds, and it also provides room to make mistakes. Beginning counselors sometimes feel they cannot err. If you believe, for example, that you can seriously damage a client with a single statement, you hold to what Meehl (1973) called "the spun-glass theory of the mind." Although counseling can be harmful, clients are not so fragile as to be psychologically destroyed by a single statement. Also, worrying about success with clients means success is unlikely (Shertzer & Stone, 1980).

 Do not assume that you know how clients react to their feelings, thoughts, and behaviors

Clients differ in their perceptions and reactions to the events in their lives and even to their psychological states. For example, one client may panic when he begins to feel depressed, believing this to be the first sign of mental illness; another may accept depression simply as a sign of physical and emotional fatigue. Be wary of assuming that how *you* respond to an event or feeling corresponds to how your *client* reacts.

Observe how clients react to their psychological states. Clients sometimes create additional problems by how they perceive the original difficulty.

> *Counselor:* So you feel angry toward your son when he deliberately disobeys you.

> *Client:* Yes.
> *(tentatively)*

> *Counselor:* I sense you're uncomfortable with that anger.

> *Client:* I get scared whenever I get mad at him. Good mothers don't get angry.

In this example, the client was troubled both by her son *and* by her anger toward him. What sometimes frightens clients about their feelings is the belief that they must *act* on every feeling. For example, if parents feel angry toward their children, they may feel like hitting the children. Typically, parents may then reject the behavior (hitting) *and* the feeling (rage, anger). Clients can learn that it is acceptable to experience any feeling but that one should not act on every feeling. Feelings are indicators of our psychological states, not the sole determinants of our behavior. We have choices.

Other clients may be afraid of or confused by their feelings. For example, persons mourning the death of a loved one may be surprised by the type and intensity of their emotions.

> *Client:* I'm just beginning to realize how much I hated my father. But I always thought I loved him!
>
> *Counselor:* You're starting to feel angry about his abuse of you as a child.
>
> *Client:* I think I'm going crazy!

Reassurance helps with such clients, as does group counseling with others having similar experiences.

> *Client 1:* This is hard for me to admit . . . but I've had really negative feelings about my father since he died.
>
> *Client 2:* What kind of feelings?
>
> *Client 1:* Well . . . anger, mostly.
>
> *Client 2:* I probably felt most angry with my father right after he died.
>
> *Client 1:* Really?
> *(surprised)*
>
> *Client 2:* Yeah . . . I don't think that's unusual at all. I know a lot of people who had to deal with that anger.

Client 1 learned that sadness is not the only feeling people can experience when someone close dies.

In sum, clients' appraisals of thoughts and feelings may result in even more intense problems (such as panic attacks) if those appraisals indicate that the feelings are wrong, are inappropriate, or fall short of imaginary standards. Help clients learn how they react to themselves.

4

Important Topics

The topics in this chapter range from major new forces in counseling to the seemingly mundane. We address important differences in counseling process and client characteristics as well as several pragmatic aspects of counseling. For many clients, the counseling process is not a matter of meeting weekly for weeks or months; crises often bring clients into counseling, and in these instances change must occur quickly. Although the counseling process is often not uniform, neither are clients. Kiesler (1971) developed the term *uniformity myth* to describe the belief that all clients are basically alike. Clients are different, and in this chapter some potentially important differences, such as gender and race, are discussed. Finally, practical issues such as the arrangement of the counseling setting, documenting the counseling progress in writing, and the referral process are briefly discussed. These details can have important effects on the counseling process.

28 Develop crisis intervention skills

Because of the stigma sometimes associated with counseling ("Only sick people see counselors"), it may take a crisis to motivate some people to seek therapy. Crises occur when individuals are faced with overwhelming problems that they feel they cannot handle (Caplan, 1961; Puryear, 1979; Slaikeu, 1990). A person in crisis could be, for example, someone who has been depressed for a long period or someone who is experiencing an acute psychotic episode. Counselors agree about basic steps for working with crisis clients. These actions differ from counselors' normal behaviors in the initial stage of counseling.

For many beginning counselors, the most frightening type of client in crisis is a person considering suicide. First, follow up on any mention of suicide or indication that clients have thought about harming themselves. Don't worry that you will be giving clients a new idea if you ask about suicide. Clients may feel relieved when asked because suicide is so difficult to discuss with friends and family.

> *Client:* I've been depressed for so long now that I've just about given up hope. I don't see any way out.

> *Counselor:* You seem so hopeless that I wonder if you have thought about hurting yourself.

A client might respond like this:

> *Client:* Oh, no, I'd never do that. But I might have to move the family away from here to look for work.

Some clients, however, have considered suicide. Counselors then determine *lethality*—that is, how likely these clients are to attempt suicide (for a more complete discussion, see Slaikeu, 1990; Sommers-Flanagan & Sommers-Flanagan, 1993).

> *Client:* Yes . . . I just want to die.

> *Counselor:* Have you thought about a way to kill yourself?

> *Client:* Yes . . .

> *Counselor:* Please tell me about it.

> *Client:* I have an empty gun in my bedroom drawer.

> *Counselor:* You've had serious thoughts about doing this. Have you been thinking about buying bullets?

> *Client:* No, not really. I just feel like I want the gun around.

The first question to ask, then, is, *Does the client have a method in mind?* The more specific, accessible, and concrete the method, the greater the likelihood the client will attempt suicide. Without knowing more than the information contained in the preceding example, we would estimate this client's lethality risk as moderate to high. A method is available, but additional action is necessary before suicide would be imminent.

The second question should elicit information about the existence and nature of previous suicide attempts.

> *Counselor:* You have a gun, but no bullets. Have you tried to commit suicide before?
>
> *Client:* Yes. I tried to slash my wrists. My wife found me and they rushed me to the emergency room.

The second question, then, is, *Have you previously attempted suicide?* The client in this example indicated that he had made a previous attempt and that it was a serious attempt. Predicting suicide, like predicting other outcomes in counseling, is problematic. However, the information about a previous, serious attempt, combined with an available method, would increase the assessment of the client's lethality to serious. In such a case, counselors ethically and legally have a responsibility to take action to preserve a client's life. This action ranges from further immediate counseling to the client's voluntary or involuntary hospitalization. Beginning counselors should learn how to assess for lethality and know what local procedures are employed with suicidal clients. Beginning counselors should *always* consult with supervisors if suicide is a possibility.

Even when suicide is unlikely, counselors working with clients in crisis take more action than they normally do. Such directiveness is embodied in the following crisis intervention strategies.

a. Take control of the situation

Particularly with suicidal clients, counselors must take direct action. In most crises, clients have given up control or perceive a great loss of control. The counselor should restore order and predictability. In other words, the counselor takes the lead.

> *Client:* Everyone here hates me! I won't leave my
> *(angrily)* room. I'm not going to eat or sleep or do
> anything but JUST SIT HERE!

> *Counselor 1:* Jim, I understand you feel you've come to the
> end of your rope. But you've been sitting on
> your bed for two days and we can't leave until
> we can find a way to help you.

> *Client:* No one else will talk to me. Why should you?

> *Counselor 2:* I care about what happens to you. Manuel
> and I are from the community mental health
> center and we'd like to listen to what's bothering you. . . .

It's worth noting that, in this example, two counselors worked as a crisis intervention team. Teams of counselors (or of counselors combined with other professionals, such as nurses and police officers) often work as crisis intervenors. A team effort decreases the stress on individual counselors, increases the potential resources available to help the client, and provides greater safety for client and counselors.

b. Determine the real client

In crisis situations, the identified client may not be the real client—that is, the person who needs attention. Each crisis situation always contains an identified client who, for example, may be threatening suicide or homicide. Identified clients, however, may be calm and clear about what they want. On the other hand, persons dealing with the identified client may be quite anxious and uncertain.

Puryear (1979) describes a crisis he encountered as a counselor in the army. He received a 2 A.M. phone call at home from the emergency room nurse, who anxiously informed him that the emergency room doctor wanted him to come to the hospital immediately. At first, the nurse would not tell Puryear why he should come to the hospital other than to say, "Dr. Smith wants you to come in now." Finally, the nurse whispered that a patient had a gun. Puryear asked that the patient be put on the telephone, learned that the patient wanted to be admitted to the psychiatric ward, and agreed to authorize his admission if the patient gave the gun to the doctor. The patient gave up the gun and Puryear authorized the admission. That resolved the crisis.

In this example, the emergency room doctor and nurse clearly were under a great deal of stress. Nurses, medical doctors, police officers, and other helping professionals with limited mental health experience sometimes become apprehensive when dealing with a crisis client. Your job as a crisis counselor, then, is to assess *all* the persons involved in the crisis situation. Treat everyone with respect, but don't accept information and commands at face value. Work with everyone who is in crisis; practically, this may mean that you will counsel the helping professional who has requested your assistance. Stay in control and avoid getting caught up in others' feelings of panic.

c. Emphasize strengths

Emphasize strengths as a way of helping clients regain control in a crisis (see Puryear, 1979, pp. 86–104, for a more complete discussion). Even a relatively minor strength or act can be noted.

> *Client:* What can I do? I'm a jerk. I'm flunking four classes and nobody will talk to me.
>
> *Counselor:* Hmmm.
>
> *Client:* I can't do anything!
>
> *Counselor:* Well, you can sit in one place for two days and not eat anything. That's a pretty difficult thing to do.
>
> *Client:* Uhmm . . . that's true.
>
> *Counselor:* Maybe you could use that toughness in some other ways.

Client: I don't know.

Counselor: Well, let's keep talking. But remember that toughness. It could be a big help.

Client: Okay.

Note the extent to which clients accept your positive attributions about them. Crisis counselors take control of crisis situations, but they give that control back at a rate consonant with the client's emotional state.

d. Mobilize social resources

Find and mobilize clients' social support network. Determine whether friends and family are available to stay with crisis clients.

Counselor: Bill, I haven't heard you mention your family in any of this.

Client: My parents don't care and neither do any of my brothers or sisters . . . except Chris. And she's in Colorado.

Counselor: Tell me about Chris.

Client: She's the person in the family I talk with the most. She convinced me to come to college because I did well in high school.

Counselor: Don't you think she might be concerned about you now, given how you feel?

Client: I don't know.

Counselor: What would you do if she felt depressed?

Client: I'd go visit her . . . maybe I could call her.

Arranging for a crisis client to stay with family or friends provides emotional stability. Once an equilibrium has been reestablished, discuss ways the client could seek further help. If possible, schedule an appointment with a counselor.

Crisis intervention provides important experience for beginning counselors. You will quickly build confidence and knowledge. Client change occurs more rapidly than it does in normal circumstances, and the difficulty level and content of crisis counseling varies greatly. At the same time, continued crisis work over an extended period can be stressful and draining to counselors (see Freudenberger, 1974). As a preventive act, many agencies rotate counselors in and out of crisis work to provide periods of less intensive effort.

29 Pay attention to issues of gender, race/ethnicity, and sexual orientation

Counselors strive to value equality between men and women while trying to understand how sources of inequality affect them (Richardson & Johnson, 1984). Although women make up the majority of clients, attention only turned to gender issues in counseling in the 1970s. Several research studies (see, for example, Broverman, Broverman, Clarkson, Rosenkrantz, & Vogel, 1970) have suggested that therapists tend to see positive qualities that are most often associated with women (such as valuing relationships) as evidence of mental illness (for example, dependency). At a minimum, female clients should receive nonsexist counseling; that is, they should receive the same type of counseling as males (Helms, 1979). Specifically, counselors should avoid fostering traditional sex roles, devaluing women, or responding to women as sex objects. At the same time, counselors should be familiar with issues that specifically impact female clients (Brown, 1994; Worell, 1992). Childhood sexual abuse, stranger and acquaintance rape, pregnancy and abortion, and eating disorders are but a few of the issues that occur more frequently for women than for men.

Counselors also need to be aware of the effects of sexism in our culture on clients (Richardson & Johnson, 1984). Sexism refers to discrimination on the basis of gender. Examples include sexual harassment and violence, media representations of women as sex objects, and sex-role stereotyping of occupations

(Abramowitz et al., 1975). Counselors should not make the mistake of attributing genuine environmental obstacles to clients' internal dysfunction (that is, problems within the client).

> *Client:* I'm having a lot of trouble finding a job. Before my husband and I divorced, I had been a wife and mother for 15 years. No one wants to hire an older woman with no skills.

> *Counselor:* It's really tough to find a job that takes advantage of the important skills you developed during your marriage. I know a program designed specifically to help women in your position find employment. I think it can help.

As have gender issues, racial and ethnic issues have been relatively neglected in counseling and psychology until recently (Carter, 1991; Casas, 1984). Counselors now strive to understand clients' behavior in the context of their cultural background (Sue, 1990a,b). Counselors should take responsibility for learning about the cultures of clients different from themselves so that they can best meet their clients' needs. For example, some Asian cultures regard direct eye contact as disrespectful and rude. A counselor who might view "poor eye contact" as indicating a problem area for a Caucasian client should not assume that it is so in an Asian American. Counselors must attend to the strengths in clients of color as well as to the difficulties resulting from living in a Euro-centric culture.

Counselors should pay attention to their stereotypes and biases about people who are different from themselves. Such misconceptions can inappropriately change the counseling process. For example, you might alter your idea of what constitutes healthy behavior on the basis of race, or you might change your judgment about the appropriateness of intensely expressed emotion on the basis of gender.

Although research in the area is controversial (see Casas, 1984), some agencies match counselors and clients by race, hoping that this assignment will facilitate counseling progress (Flaskerud, 1990; Pope, 1979). The assumption is that the greater the similarity between counselor and client (in terms of race, sex, culture, and so on), the more easily a therapeutic relationship will develop. However, generalizations about group status or gender should be applied cautiously to counseling practice because of the large individual differences that exist within these groups.

Finally, counselors should not assume that a client's sexual orientation is heterosexual. Whereas some gay and lesbian clients are open about their sexuality, some may not feel safe sharing this with you. Others are confused and still struggling to identify or accept their sexual orientation (see Clunis & Green, 1988).

Successful counseling depends largely upon the counselor's awareness of his or her feelings about and comfort with sexuality. A counselor who considers lesbians or gay men "sick" or "perverted" must work through these biases before attempting to counsel this group of clients. Counselors must also be sensitive to the social, legal, and economic barriers gays and lesbians face in our society. These obstacles present special difficulties that others in our society do not face.

> *Client:* I'm really in love with my partner and I would like her to meet my family, but I'm afraid of what they will think about me.

> *Counselor:* You're worried that they won't accept you if you tell them you're a lesbian.

> *Client:* That's right. My father makes nasty remarks whenever the subject of gays comes up. He thinks it's immoral. But I'm tired of lying to them.

> *Counselor:* This is a very important issue for you. Let's explore it some more.

 Be open to group and family approaches

Many counselors see one-to-one counseling as the treatment of choice for all clients. However, clinical experience and some research suggest that substantial advantages exist for group and family approaches (see Bednar & Kaul, 1978; Gurman & Kniskern, 1978; Yalom, 1995).

Particularly from an administrative view, groups are very efficient. If one counselor can work with ten clients in one 90-minute session, that represents considerable savings in comparison to one counselor working ten 1-hour sessions. From a client's perspective, groups provide emotional support, models of coping behavior, and evidence that one's problems are shared by others (for elaboration of these and other advantages of groups, see Schneider Corey & Corey, 1992; Yalom, 1995).

Selecting clients by similar age or social maturity can produce a homogeneous group that grows in cohesion and helpfulness (Shertzer & Stone, 1980). Members of helpful groups feel part of the group as a whole and benefit from the open giving and receiving of feelings and ideas (see Barrett-Lennard, 1974; Schneider Corey & Corey, 1992).

> *Client 1:* I want to be an engineer, but I really don't know if I can do it.
>
> *Client 2:* I felt the same way last year. People thought I was odd because I didn't want to be a secretary.
>
> *Client 1:* That's right. That's what my parents say.

Counselors normally select clients for groups through individual interviews. In this way, counselors can determine if clients possess sufficient social skills and emotional stability to benefit from a group.

In family therapy, counselors often employ a systems approach. Family therapists see individuals' problems as relating to the social systems to which they belong, the most important of which is the family (deShazer, 1982; Minuchin, 1974). Families often act to preserve their status quo; in some families, this means that members act to keep certain individuals "sick." The first task of the family counselor, then, is to observe the communication patterns of family members (see Satir, 1988). Next, the counselor points out the maladaptive interactions and helps the family members change so that a new structure of communication and interaction takes hold.

Refer carefully

You cannot help every client. You may lack the necessary skills; clients may move to a distant locale; you and a particular client may simply be unable to work together. Regardless of the reason, you should be aware of other resources in the community.

> *Client:* We're moving to Seattle in a month, but I'd still like to continue therapy with someone.
>
> *Counselor:* Actually, I know several counselors in that area who work as private practitioners. I can give you their names and phone numbers if you like.
>
> *Client:* That would help. Would you be willing to speak to my new counselor if that's necessary?
>
> *Counselor:* Yes. You should sign a release of information form so I can do that.

To respect clients' rights and to ensure that they follow through, explain the reason when you make a referral. After your explanation, seek a reaction to your referral.

> *Counselor:* I think a psychological test might save time and help me get a better sense of what is happening with you.
>
> *Client:* What kind of test?
>
> *Counselor:* It's called the Minnesota Multiphasic Personality Inventory, or MMPI. It takes about an hour and a half, and it will help us decide what direction to go in counseling. What do you think?
>
> *Client:* That's okay. I think maybe I took that test a few years ago at the school I transferred from.
>
> *Counselor:* Okay. We have to set up an appointment for you at the testing center. Maybe we can also track down those previous results—with your permission.

Moxley (1989) employs the term *linkage* to describe strategies that are employed by mental health professionals to connect clients with needed services through the referral process. To ensure a connection between a client and a new agency, for example, the counselor must first address any fears or misconceptions the client holds about the agency. The referring counselor should also attempt, with the client's written permission, to transmit essential information about the client to the agency. Such information might include the purpose of the referral, specific needs of the client, and relevant assessment data. Moxley also suggests that counselors telephone clients after the referral to be certain that the client did connect with the agency and is receiving adequate services.

 ## 32   Watch for deterioration in clients

Counselors rarely consider that clients may be harmed by the counseling process. But considerable evidence exists to suggest that some clients deteriorate in counseling (Bergin & Lambert, 1978; Mohr, 1995).

In research on groups, Lieberman, Yalom, and Miles (1973) found that leaders who were frequently aggressive, authoritarian, and inappropriately self-disclosing produced the greatest number of dropouts and negative changers (people who became worse after counseling). Stuart (1974) attempted to educate 935 students in the seventh and ninth grades through ten classroom sessions about drugs. These students reported significant increases in their knowledge about drugs—and in their use of alcohol, marijuana, and LSD.

No generic prescriptions can be given for avoiding negative change in counseling. However, you should assess and observe client reaction to every intervention. The following interaction, for example, might occur in the beginning stages of counseling.

> *Client:* I have faint recollections about my father hurting me when I was very young . . .

> *Counselor:* Did your father abuse you?

Client: I . . . I don't know. I'm starting to get very upset.

Counselor: You seem anxious . . .

Client: Yes. I'm very scared.

Counselor: How do you feel about talking about this?

Client: I'm not ready. Not yet . . . at least not this fast.

Counselor: That's okay. We'll come back to this later.

In this example, the counselor assessed the client's feelings about
further probing of the potential abuse. Since the counseling
process was still in the early stages, the counselor decided to
table the probe and reapproach the issue later in counseling.

Pushing clients too fast may be harmful. Some counseling approaches, such as Gestalt, emphasize frustrating and confronting
clients. Gestalt counselors, however, still observe their clients to
determine the effects of confrontation.

 Establish an interest in counseling research

Counseling is an art in the throes of becoming a science. At present, practitioners can do more than researchers can explain.
For example, counselors can help clients with phobias and depression, but researchers can only partially explain the etiolo-
gy of those problems and why counseling works with them
when it does. A substantial split separates practitioners and researchers: researchers dismiss practitioners as "touchy-feely,"
whereas practitioners view current research as irrelevant (Gelso,
1979; Meehl, 1956). However, practitioner-scientists, theoreti-
cians, and other researchers are the groups likely to discover
the patterns and order that occur in effective counseling and
psychotherapy.

Competent counselors and psychotherapists should understand the research process, if not contribute to it (see Heppner,
Kivlighan, & Wampold, 1992). Keep up with current research by
reading such journals as these:

American Journal of Psychiatry
Behavior Therapy
Clinical Psychology Review
Cognitive Therapy and Research
Journal of Consulting and Clinical Psychology
Journal of Counseling and Development
Journal of Counseling Psychology
Professional Psychology: Research and Practice

Counselors who understand the research process are better able to conduct, cooperate in, and coordinate research in applied settings (such as hospitals, clinics, college counseling centers, and community mental health centers). Applied settings, because of their primary orientation to service, typically present counseling researchers with substantial obstacles to establishing and maintaining research, be it single case designs or large outcome studies. Conducting research in field settings such as community mental health agencies, however, provides an excellent source of research questions and offers great potential for producing results relevant to actual clients and problems.

Beginning counselors who desire to learn about research should ally themselves with a faculty member or counselor who is conducting empirical work (Gelso, 1979). Such an alliance teaches the beginner research skills and demonstrates the day-to-day difficulties of conducting counseling research. Researchers in applied settings must learn, for example, how to protect their scarce research time against service demands and how to handle the politics of conducting research in a service agency (Gelso, 1979).

 ## Document your work

After each counseling session, take notes. Documentation has five benefits: (1) records provide support for you in any legal action; (2) records allow you to accurately inform your supervisor of process and issues; (3) records enable you to comply with agency standards for accountability (such as the number of hours spent in individual counseling or issues to be addressed should a client terminate and later return for further counseling); (4) records remind you of treatment history and amount of progress; and (5) records provide the basis for financial reimbursement by third-party payers (for example, insurance companies) and for charging fees to clients.

What should you write in your notes? This depends partially on where you work, but in general you should briefly describe for each session (1) principal content and themes; (2) noteworthy counselor interventions, with client response; (3) client status (including descriptions of client cognitions, affect, and behavior); (4) expected client behavior during the period leading up to the next session, including any homework; (5) expected time and interventions still required to meet counseling goals; and (6) indications of client expectations of the outcome of counseling. Your notes could also include tests administered (with results and client reactions), diagnoses, your reaction to the session, questions to ask supervisors or colleagues, and relevant client interactions with family and friends.

Such notes should be kept under lock and key where no one can access them purposefully or by accident. Computerized record-keeping systems are available and also must have restricted access.

 ## Persevere with no-shows

Beginning counselors sometimes have difficulty knowing how to react when their clients fail to attend scheduled sessions. One student counselor had his first three clients fail to show up, and the student assumed that something was wrong with his initial phone contact. Another counselor became angry with a client for failing to attend a session; he subsequently decided not to attend the next session, rationalizing that he wasn't certain the client was going to attend. The client did.

Avoid judgments on no-shows. At the next contact (a follow-up phone call or scheduled session), ask the client about the absence. Determine if the failure to attend reflects on progress (or lack thereof) in counseling or the client's characteristic manner of dealing with people. Also determine if the client conceives of time in different ways than you do. If helpful, provide a written reminder, such as an appointment card and a rescheduling method at the end of each session. With special needs clients, build in some flexibility.

In essence, regard no-shows as material for therapy rather than as a personal affront.

Counselor: Amy, you missed the last two sessions.

Client: Yeah. I just didn't feel like coming.

Counselor: Tell me more.

Client: After our last session I felt so depressed I didn't want to talk about my problems anymore.

No-shows become especially complicated when clients interpret them as additional failures. For example, clients may miss a meeting after a particularly difficult session. However, they may then see the no-show as an indication that they are unable or unworthy to continue counseling. Premature termination becomes a real possibility. You might avoid this failure loop by giving such clients excused absences—that is, planned permission to miss a session. This may increase some clients' willingness to attend future meetings following a no-show.

Client: I missed the first session because I didn't feel like talking. And then I got embarrassed because I didn't show up.

Counselor: So you missed the second session because of your embarrassment. Why don't we just agree it's okay for you to miss a session if you call ahead and cancel?

Client: Gee . . . that sounds okay.

If *you* must cancel a session, inform the client as soon as possible and reschedule.

Arrange the physical setting appropriately

The counseling process is affected by the physical characteristics of the counselor and the counseling setting (see Langs, 1973, and Sommers-Flanagan & Sommers-Flanagan, 1993, for more complete treatments).

a. Dress appropriately

Dressing for success may be more necessary in business than in counseling, but some clients judge counselors by their attire. An appearance that matches that of other colleagues is fitting. Dress becomes problematic when counselors draw attention to themselves through apparel.

b. Attend to physical space

Observe how clients use space. Some clients may prefer that there be no physical objects between you, whereas others may feel safer with a barrier (such as a desk or small table). We prefer to sit facing clients, although some counselors have suggested that sitting at an angle increases client comfort. If possible, provide comfortable chairs for counselor and client. This makes it easier for both to concentrate during the entire length of the session.

c. Conduct counseling in a quiet setting

If you can overhear another conversation during a counseling session, it's likely that someone can overhear your (confidential?) discussion. Respecting clients' privacy and confidentiality means counseling should occur in a quiet, private setting.

d. Avoid interruptions and distractions

Don't accept phone calls; don't read the mail; don't finish your lunch. Attention in a counseling session belongs exclusively to the client, except in the case of an emergency.

e. Be prompt

If clients arrive late or leave early, that's an issue to address in counseling. In any event, clients should not be wondering about the whereabouts of their counselors ten minutes after the scheduled start. A regularity to the beginning and end of counseling sessions signals that clients can trust this relationship. Similarly, arranging the session for the same time each week or day can signify the counselor's commitment to the relationship.

Some clients measure their importance to the counselor by how often their sessions exceed the scheduled time. Be explicit with clients about the length of a session. A typical individual counseling session lasts 45 or 50 minutes. This is enough time to go beyond the superficial, but not so long that client and counselor are exhausted.

f. Invest in a box of tissues

Sometimes clients cry. It's generally not a good idea, however, for the counselor to hand a tissue to a client. It may imply, to some clients, that it is time to stop being sad. Let clients use tissues when they're ready.

g. Remember confidentiality

Restrict your communications about clients to professional colleagues in professional settings.

Counselors may be tempted to discuss clients in public places such as restaurants, teachers' rooms, or informal gatherings. Particularly in small towns and other settings where people who overhear those conversations might know the individuals involved, talking about clients in public has the potential for serious trouble. Restrict your discussion of clients to supervisors and necessary colleagues in work settings.

 Learn how to conceptualize clients

Creating a visual model of the counseling process and outcome can be a useful method for understanding the complexities of working with clients. Models are graphical representations of important elements influencing the behavior of a particular client. All counselors develop ideas about what is troubling their clients and what they could be doing to help them (Teyber, 1992). However, these mental models are often implicit and, occasionally, ineffective. For example, new counselors often act as if the best way to intervene with clients is to give advice. This model—that advice translates into problem resolution—is typically an inaccurate depiction of the needed counseling process.

By closely observing clients and ourselves, we can improve our counseling. Close observation in this context refers to (1) representing our experiences with and ideas about clients in terms of models and then (2) assessing the efficacy of these models. Suppose, for example, that Robin, a high school student, begins to have frequent fights at school after she learns that her parents are considering a divorce. When Robin gets into a fight, several outcomes occur, as shown here.

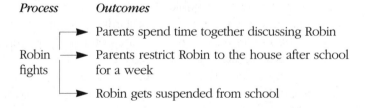

Process   *Outcomes*

Robin → Parents spend time together discussing Robin

fights → Parents restrict Robin to the house after school for a week

→ Robin gets suspended from school

Note that when Robin gets in trouble at school, several events occur that differ in their attractiveness to Robin. For example, Robin may be ashamed at being suspended and angry about being restricted but secretly pleased that her arguing parents must spend time together to decide what to do about her. Although this is a simplified conception of what may be happening with Robin and her family, it is a useful starting point for considering how best to help Robin.

Given this knowledge, we could represent the causes of Robin's problem as:

Parents argue;   Robin gets   Parents spend time
discuss divorce ⟶ in trouble   ⟶   discussing Robin

This model represents our initial ideas about the causes, origins, or development of Robin's problem. We can also include in the model the procedures we believe are necessary for the resolution of Robin's problem:

Parents argue;   Robin gets
discuss divorce ⟶ in trouble

                     Parents spend time
                     discussing Robin   ⟶ Strengthen parents'
                                            relationship with
                                            couples counseling

Thus, our initial guess about how to help Robin involves a family systems approach. That is, we speculate that the source of Robin's problems lies in the family, not in Robin herself (Bowen, 1978; see also Chapter 6). We suspect that if we strengthen the parents' relationship or the family as a whole, Robin will get something she needs and this will decrease the frequency of her getting in trouble at school.

Again, this conceptualization is just a starting point in a treatment plan. The model may turn out to be completely wrong, very helpful, or, most likely, in need of some modification. The importance of a model is that it can serve as one source of information about how to proceed in counseling. Although space does not permit a detailed explanation, the next steps would involve collecting qualitative and perhaps quantitative data that would enable the counselor to chart trends in process and outcome variables. These data would provide documentation about counseling outcomes as well as indicate modifications to the model that could result in more effective interventions (see Hartman, 1984; Paul, 1986).

Counselor, Know Thyself

In no other profession does the personality and behavior of the professional make such a difference as it does in counseling. Beginning counselors need to work at increasing their self-awareness as well as their knowledge of counseling procedures. Your willingness to be open to supervision, to accept clients' failures and criticisms, to participate in counseling yourself when appropriate, and to acknowledge your limits will contribute to your eventual success and satisfaction.

 ## 38 Become aware of your personal issues

Every beginning counselor will eventually confront difficult questions about her or his personal issues. Answers to these questions may not be readily apparent. Your issues, however, do influence how you counsel (see Egan, 1990).

Counseling others and being counseled, having competent supervision, and developing a theory of counseling will help you answer the following questions.

a. How did you decide to become a counselor?

Many counselors became interested in a helping profession because family and friends sought them out to listen and provide help. Other counselors were once clients themselves and decided to follow in the footsteps of counselors who helped them.

Your reasons for becoming a counselor may affect how you counsel. For example, did your assistance to friends include acting as a rescuer? Rescuers are persons who do the problem solving and helping in their social groups. As a counselor, will you again rescue?

> *Client:* I just don't have the willpower to eat more.
>
> *Counselor:* You can call me any time of the day or night when you need help.

Can you allow clients to handle their own problems?

> *Client:* I just don't have as much willpower as I once had.
>
> *Counselor:* What stops you from using the willpower you do possess?

A counselor can solve a problem for a client. A counselor can also help clients learn how to solve problems successfully. The difference is important.

b. With what emotions are you uncomfortable?

Clients often experience intense feelings such as rage or grief. Because of their lack of exposure, beginning counselors may be uncomfortable with some feelings, at least at the intensity level expressed by clients. Will you allow clients to express those emotions? Are there feelings that you avoid, that you will steer your clients away from?

> *Client:* I just hurt so much since the divorce . . .
>
> *Counselor:* Try not to think about it. Are you having any luck getting dates?

In this example, the counselor unconsciously moved the discussion away from the intense hurt. Counselor training includes exposure and desensitization to a range and intensity of feelings new to beginning counselors. Identify the issues that make you uncomfortable: for example, as a new counselor one of the authors was afraid of talking with clients about their sexuality (particularly homosexuality), their feelings of rage, and their feelings of helplessness. Without help from supervisors and the experience of personal counseling, the author would have continued to avoid those areas when clients ventured near them.

c. What amount of progress is acceptable?

Although individuals differ in the progress they make in counseling, beginning counselors are sometimes surprised at how slowly clients change. Most clients cannot alter their troublesome behavior quickly or at the instruction of the counselor. For counselors, the issue becomes how much progress is acceptable.

> *Client:* I know I agreed to speak in each of my classes, but . . .

> *Counselor:* (silence)

> *Client:* . . . but I only asked one question, in physics.

What will be a blow to your confidence as a counselor? A client who drops out? Ask yourself what rewards you expect as a counselor. Do you want money, status in the community, or intimacy without threat?

d. How will you deal with your clients' feelings for you?

If the counseling relationship becomes at all significant to the client, she or he will have feelings for you. Interestingly, these feelings may have little to do with what you've said or done. Clients may transfer feelings from past relationships onto their perceptions of *you*. Clients' feelings about their counselors are generally referred to as *transference* (see Hansen, Rossberg, & Cramer, 1994, for a general introduction; see Kahn, 1990, and Robertiello & Schoenewolf, 1987, for advanced examples). How will you feel if a client perceives you as attractive, wise, or racist?

> *Client:* I know you don't like me.

Counselor: What makes you say that?

Client: You don't like me because I'm blind!

In general, counselors process such feelings with their clients. Because this processing may influence progress, it is important to develop your ability to recognize and work with clients' feelings for you.

e. *How will you handle your feelings for your clients?*

You may have strong or ambivalent feelings about clients. Counselors' feelings about their clients are generally referred to as *countertransference* (again, see Hansen et al., 1994; Kahn, 1990; Robertiello & Schoenewolf, 1987). Is it okay for you to feel sexually attracted to a client? What will you do with that feeling? Perhaps you came from a family in which you were abused as a child. Will you be able to work with child abusers? You too may transfer your feelings about past relationships onto your clients.

Client: I've always gotten my way. I was always the biggest kid in grade school, and I just bullied anybody who got in my way.

Counselor: (angrily) Don't you think you were a jerk to do that?

Counselors-in-training (and professional counselors, for that matter) often engage in their own therapy both for personal growth and to be better able to help their clients.

Client: I've always gotten my way. I was always the biggest kid in grade school, and I just bullied anybody who got in my way.

Counselor: Anyone who gets in your way gets pushed aside.

Client: Yeah . . . but bullying doesn't always get me what I want with my family and coworkers.

On the other hand, the relationship between counselor and client sometimes parallels other relationships. For example, if sarcasm is part of a client's interpersonal style, sarcasm is likely to be part of the client's verbal behavior with the counselor. What you as the counselor feel about the client, as a person, may also reflect what other people feel. Some counselors judge progress in counseling (particularly for clients with interpersonal difficulties) by changes in the relationship between counselor and client.

> *Counselor:* You know, John, I sense that you've felt more relaxed talking with me the past couple of weeks.
>
> *Client:* I guess I don't feel quite as suspicious of people as I used to. Funny—a friend of mine said the same thing last week.

In this example, the counselor noticed less tension in the relationship. The client acknowledged a change in his interpersonal style both in and out of counseling.

f. Can you be flexible, accepting, and gentle?

The attitudes you hold toward counseling and clients affect the process and outcome of counseling. Flexibility refers to the counselor's ability to be creative, open, and aware of the here-and-now in counseling (Hansen et al., 1994; Van Kaam, 1966). Acceptance is the counselor's willingness to hear and understand whatever the client says without judgment, without conditions. Gentleness is the capacity to be kind and considerate even when clients are abrupt, afraid, and defensive.

> *Client:* You're not helping me. I can't talk about this!
>
> *Counselor:* How difficult it is for you to remember your
> *(softly)* childhood . . .
>
> *Client:* It hurts so much.
>
> *Counselor:* *(silence)*

As a profession, counseling has members who are probably more accepting of individual differences than professionals of any other group. Such tolerance can be learned.

 ## 39   Be open to supervision

The preceding set of questions may be overwhelming to beginners, who may wonder how they will ever become as self-aware. Relax—no counselor is expected to be self-actualized. You are expected, however, to know the areas where you have personal difficulties. Counseling involves knowing yourself, not just knowing techniques and theory; in this area, a competent supervisor can be invaluable. Supervision is more than instruction, consultation, or direction (see Whiteley, 1982; Bartlett, Goodyear, & Bradley, 1983). Supervision resembles counseling when it involves exploring the supervisee's personal issues.

Counselor: So when he started to talk about dying, I really became anxious. I started to sweat and feel really dizzy.

Supervisor: What was it that frightened you?

Counselor: I don't know . . .

Supervisor: It was when he mentioned dying that you became anxious.

Counselor: I just haven't dealt with death much, and he was feeling so afraid. That triggered the same feelings in me.

Supervision is the best setting in which to explore feelings about your clients. Of course, beginning counselors may find it useful to enter counseling themselves. Supervision cannot be entirely concerned with the supervisee's issues, and personal counseling may facilitate the new counselor's growth as a therapist. You also gain the perspective of seeing what it's like to be on the other side of the counselor/client relationship.

We consider supervision so important that we strongly recommend that you continue supervision after your official training ends. Maintaining the discipline and perspective provided by good individual or group supervision seems essential to consistent, high-quality counseling.

 Don't hide behind testing

Because clients' expectations of counselors' expertise can influence the success of counseling (see Strong, 1968), beginning counselors may be tempted to boost their credibility through testing.

> *Counselor:* Your vocational test results indicate that you
> *(authorita-* should be a salesperson.
> *tively)*
>
> *Client:* Really?
>
> *Counselor:* No doubt about it—look at these scores.

In this example, the counselor implies that the test is a foolproof method of selecting a career. Although testing data may be helpful in understanding clients and in designing counseling strategies, the counselor should remember to explain the limits of the test along with the results (American Psychological Association, 1992). This is particularly important when tests are administered by computer; clients tend to view computer-administered instruments as more scientific (Herr & Best, 1984).

Counselor credibility comes with the confidence produced by experience and knowledge, not props. Being an expert does not mean that you are dogmatic or authoritarian (Cormier & Cormier, 1991).

> *Client:* So which one of these careers on the computer
> list should I choose?
>
> *Counselor:* Well, this inventory wasn't designed to make a
> choice for you. This list should help you get an

idea of what kinds of careers might be best for you to explore.

Client: Oh.

Tests are tools. They are imperfect and useful at the same time. Test results, in and of themselves, never dictate action. Counselors employ tests as one more technique in their repertoire for helping clients.

 ## On ethical questions, consult

Counselors inevitably find themselves in gray zones. Issues of counselors' duty to warn, dual relationships, confidentiality and privacy, client rights, and personal relationships with clients are subjects with considerable potential for ambiguity and risk (Corey & Schneider Corey, 1993). In addition to knowing the formal ethical standards of your profession (see American Association for Counseling and Development, 1988; American Psychological Association, 1992), you must be willing to discuss ethical issues with supervisors and colleagues.

Counselor 1: He started to talk about wanting to hurt his ex-wife.

Counselor 2: Did he mention a method?

Counselor 1: I didn't think to ask about that. I'm wondering if I have to notify her about his threat.

States vary in the specifics of law, but the Tarasoff case (see Schmidt & Meara, 1984; Stanard & Hazler, 1995) set the precedent that counselors have a legal responsibility to notify the persons whom their clients threaten. For example, if you were counseling a divorced man who threatened to kill his ex-wife, you would be legally obligated to contact her and warn her about the threat. Obviously, such a warning breaks confidentiality, so counselors must carefully assess the seriousness of threats made by clients. In general, you also have the obligation to protect clients' lives if they are suicidal and to report child abuse if you learn of it during counseling. Some agencies and counselors provide clients with a written form describing these limits to confidentiality, along with relevant information about clients' rights to terminate treatment and review records (Schmidt & Meara, 1984).

New counselors also express concern about the limits of their skills. Counselors refer clients whom they cannot help. New counselors may have difficulty deciding if they have the necessary skills. Consult with your supervisor when you begin to question whether you can help a particular client. Your supervisor may teach you new skills or help you find other assistance.

On ethical questions, don't get caught out on a limb. Always consult if you have any doubt. Colleagues and supervisors may be more objective and able to help you decide on an appropriate course of action.

6

A Brief Introduction to Intervention

Many counselors see therapy primarily as a relationship-building endeavor, but most counselors emphasize the necessity of further intervention for change to occur (Egan, 1990). How to intervene effectively divides today's counselors and psychotherapists.

Counselors traditionally have discussed people in terms of three systems: affect, cognition, and behavior. *Affect* refers to the feelings we experience and express (such as anger or sadness); *cognitions* are the thoughts we think (such as, "It's not my fault I flunked the test"); and *behaviors* are the overt acts we do (such as smoke three packs of cigarettes a day). Clients frequently report that they seek counseling to change such painful feelings as anxiety or depression. Counselors and psychotherapists with different orientations disagree vigorously about which of the three systems should be targeted for intervention. The task becomes further complicated by controversy about permanence of change (Mahoney, 1987) and the desynchrony of the systems (Hodgson & Rachman, 1974; Rachman & Hodgson, 1974). *Permanence of change* relates to the question of symptom substitution: if we change behavior without attention to underlying feelings or beliefs, will the problem resurface elsewhere? *Desynchrony* refers to the commonly observed lack of correspondence among measures of individuals' affect, cognitions, and behaviors. For example, even if clients change their behavior, weeks may pass before the corresponding feeling change occurs.

Research findings support the general efficacy of counseling, indicating that the average client who receives treatment enjoys more improvement than do two-thirds of persons who do not receive counseling (Landman & Dawes, 1982; Smith & Glass, 1977). On the other hand, reviews of outcome studies generally negate claims of overwhelming superiority for any one approach. Counselors still grapple with the question of which approach or counselor works best with which client or problem (cf. Krumboltz, 1966, with Glidden-Tracey & Wagner, 1995). Rather than fight battles over which approach is best, researchers have become increasingly interested in discerning common elements in different counseling approaches and theories (see Ivey, 1980, and Staats, 1983, for a fuller discussion).

Beginning counselors should become familiar with the basic theory and practice of many approaches. Only then can you make the informed choices necessary to create, integrate, and structure your personal method. The following list contains basic counseling texts and references for 10 well-known counseling approaches and provides a reasonable starting point for exploring the field.

Basic counseling texts

The following texts present overviews of various counseling approaches, counseling professions, and basic counseling skills. Elaboration of the content of *The Elements of Counseling* can be found in these books.

References

Corsini, R., & Wedding, D. (1995). *Current psychotherapies* (5th ed.). Itasca, IL: F. E. Peacock.

Egan, G. (1994). *The skilled helper: A problem-management approach to helping* (5th ed.). Pacific Grove, CA: Brooks/Cole.

Hansen, J., Stevic, R., & Warner, R. (1992). *Counseling: Theory and process.* Boston: Allyn & Bacon.

Ivey, A. (1994). *Intentional interviewing and counseling* (3rd ed.). Pacific Grove, CA: Brooks/Cole.

Osipow, S., Walsh, W. B., & Tosi, D. (1984). *A survey of counseling methods.* Homewood, IL: Dorsey Press.

Shertzer, B., & Stone, S. (1980). *Fundamentals of counseling* (3rd ed.). Boston: Houghton Mifflin.

Waldinger, R. (1986). *Fundamentals of psychiatry.* Washington, DC: American Psychiatric Press.

Person-centered counseling

Person-centered, client-centered, or nondirective therapy centers on the work of Carl Rogers. This approach is most unique in its emphasis on clients' ability to determine relevant issues and to solve their problems. Person-centered counselors tend to see their clients positively (since all people are assumed to be striving for self-actualization) and to respond to clients with warmth, support, unconditional positive regard, genuineness, and empathy. Client-centered counselors assist clients in the change process by focusing on congruence and affect. The counselor notices client feelings and empathizes with those feelings to help clients fully experience their affect and become more open to their life experiences.

References

Kahn, M. (1990). *Between therapist and client.* New York: W. H. Freeman.

Meador, B. D., & Rogers, C. (1979). Person-centered therapy. In R. Corsini (Ed.), *Current psychotherapies* (2nd ed.). Itasca, IL: F. E. Peacock.

Patterson, C. H. (1985). *The therapeutic relationship: Foundations for an eclectic psychotherapy.* Pacific Grove, CA: Brooks/Cole.

Rogers, C. (1951). *Client-centered therapy.* Boston: Houghton Mifflin.

Rogers, C. (1986). Client-centered therapy. In I. Kutash & A. Wolk (Eds.), *Psychotherapist's casebook: Theory and technique in practice.* San Francisco: Jossey-Bass.

Behavioral counseling

Behavioral counselors tend to be the pragmatists of the counseling profession. After all, they maintain, if it is behavior that we ultimately want to change (be it smoking, anxiety about school performance, or depression), then it is behavior that we should target in counseling. Behavioral counselors focus on inappropriate learning as the source of client problems. Thus, clients may have inappropriately learned to associate social situations with anxiety, and consequently a regimen of relaxation and assertiveness training is prescribed. Behavioral counselors pay attention to reinforcement as provided in clients' environments; if a child misbehaves at home, a behavioral counselor might teach parents about how to reward that child for more appropriate behavior.

References

Agras, W., Kazdin, A., & Wilson, G. (1979). *Behavior therapy.* San Francisco: W. H. Freeman.

Krumboltz, J., & Thoresen, C. (1976). *Counseling methods.* New York: Holt, Rinehart & Winston.

Masters, J., Burish, T., Hollon, S., & Rimm, D. (1987). *Behavior therapy: Techniques and empirical findings* (3rd ed.). San Diego, CA: Harcourt Brace Jovanovich.

Skinner, B. F. (1971). *Beyond freedom and dignity.* New York: Knopf.

Wilson, G., & Franks, C. (Eds.). (1982). *Contemporary behavior therapy.* New York: Guilford.

Wolpe, J. (1990). *The practice of behavior therapy* (4th ed.). New York: Pergamon Press.

Cognitive, cognitive/behavioral, and social learning counseling

Counselors with a cognitive orientation represent the latest movement in the counseling profession. In one form or another, these counselors consider inappropriate thoughts to be the cause of painful feelings and harmful behavior. Counselors like Ellis view irrational beliefs (beliefs without empirical evidence) as the target for interventions, whereas Beck describes how selective attention, magnifying problems, and illogical reasoning can lead to depression. Cognitive and cognitive/behavioral counseling grew from the behavioral counseling movement and share a tradition of respect for applying research to practice and doing counseling research.

Social learning counselors tend to focus specifically on individuals' expectations. As a result of their social experiences, individuals learn to expect that (1) some events are more personally rewarding than others; (2) certain behaviors can produce desired events, although there may be events that are uncontrollable; and (3) people differ in their feelings of competence for doing the behaviors that can produce desired events. Although social learning theorists view expectations as the driving force, these counselors often do not directly intercede with cognitions. Instead, they modify clients' expectations through actual performance (such as in a gradual exposure to increasingly fearful situations, as is done with snake phobics) or through use of models who demonstrate skilled behavior.

References

Bandura, A. (1969). *Principles of behavior modification*. New York: Holt, Rinehart & Winston.

Bandura, A. (1977a). Self-efficacy theory: Toward a unifying view of behavioral change. *Psychological Review, 84,* 191–215.

Bandura, A. (1977b). *Social learning theory*. Englewood Cliffs, NJ: Prentice-Hall.

Beck, A. (1976). *Cognitive therapies and the emotional disorders*. New York: International Universities Press.

Cormier, W., & Cormier, L. (1991). *Interviewing strategies for helpers: Fundamental skills and cognitive behavioral interventions* (3rd ed.). Pacific Grove, CA: Brooks/Cole.

Ellis, A., & Grieger, R. (1977). *Handbook of rational-emotive therapy*. New York: Springer.

Ellis, A., & Harper, R. (1976). *A new guide to rational living*. North Hollywood, CA: Wilshire Book Company.

Emery, G., Hollon, S., & Bedrosian, R. (Eds.). (1981). *New directions in cognitive therapy*. New York: Guilford.

Hollon, S. D., & Beck, A. T. (1994). Cognitive and cognitive-behavioral therapies. In A. E. Bergin & S. L. Garfield (Eds.), *Handbook of psychotherapy and behavior change* (4th ed., pp. 428–467). New York: Wiley.

Kelly, G. (1955). *The psychology of personal constructs.* New York: Norton.

Mahoney, M. J., & Arnkoff, D. (1978). Cognitive and self-control therapies. In S. Garfield & A. Bergin (Eds.), *Handbook of psychotherapy and behavior change: An empirical analysis.* New York: Wiley.

Mahoney, M. J., & Lyddon, W. J. (1988). Recent developments in cognitive approaches to counseling and psychotherapy. *The Counseling Psychologist, 16,* 190-234.

Meichenbaum, D. (1977). *Cognitive behavior modification: An integrative approach.* New York: Plenum.

Meichenbaum, D. (1985). *Stress-inoculation training.* New York: Pergamon Press.

Mischel, W. (1986). *Introduction to personality* (4th ed.). New York: Holt, Rinehart & Winston.

Rosenthal, T., & Zimmerman, B. (1978). *Social learning and cognition.* New York: Academic Press.

Rotter, J. (1954). *Social learning and clinical psychology.* Englewood Cliffs, NJ: Prentice-Hall.

Rotter, J. (1971). *Clinical psychology* (2nd ed.). Englewood Cliffs, NJ: Prentice-Hall.

Rotter, J. (1982). *The development and applications of social learning theory: Selected papers.* New York: Praeger.

Rotter, J., Chance, J., & Phares, E. (1972). *Applications of a social learning theory of personality.* New York: Holt, Rinehart & Winston.

Williamson, E. (1959). Some issues underlying counseling theory and practice. In W. Dugan (Ed.), *Counseling points of view.* Minneapolis: University of Minnesota Press.

 ## Gestalt counseling

Fritz Perls is the counselor most strongly associated with Gestalt counseling. Perls frustrated clients to help them move toward self-support and away from therapist support. Gestalt counselors also emphasize body movement as a method of experiencing feelings and facilitating psychological growth. As do person-centered counselors, Gestalt counselors pay particular attention to noticing client feelings, staying in the here-and-now, and avoiding intellectual analysis of problems.

References

Downing, J. (1976). *Gestalt awareness.* New York: Harper & Row.

Hardy, R. E. (1991). *Gestalt psychotherapy.* Springfield, IL: Charles C. Thomas.

Perls, F. (1970). *In and out of the garbage can.* New York: Bantam Books.

Perls, F. (1971). *Gestalt therapy verbatim.* New York: Bantam Books.

Perls, F., Hefferline, R., & Goodman, P. (1977). *Gestalt therapy: Excitement and growth in the human personality.* New York: Bantam Books.

Polster, E., & Polster, M. (1973). *Gestalt therapy integrated.* New York: Brunner/Mazel.

Zinker, J. (1977). *Creative process in Gestalt therapy.* New York: Random House.

 ## Transactional analysis

One of the strengths of transactional analysis (TA) is its straightforward view of personality and interpersonal interaction. TA counselors suggest that each individual is composed of three parts: Parent, Adult, and Child. Our Parent tells us what is right; our Adult makes decisions and tests reality; and our Child plays and has needs. Part of TA counseling involves teaching this personality theory, which is easy for clients to learn and remember. Conflict among Parent, Adult, and Child explains both intrapsychic and interpersonal difficulties. For example, adults often find that their parents still view them as children and are incapable of communicating adult-to-adult with them. TA counselors help their clients achieve a balance among their three parts, with a slight emphasis on the decision-making Adult. A person with too much Child would likely have difficulty living a responsible life; a person with too much Parent might have difficulty enjoying life.

References

Berne, E. (1986). *Transactional analysis in psychotherapy.* New York: Ballantine.

Goulding, R., & Goulding, M. (1978). *The power is in the patient.* San Francisco: TA Press.

Harris, T. (1982). *I'm OK—You're OK.* New York: Avon.

Jongeward, D., & Scott, D. (1976). *Women as winners: Transactional analysis for personal growth.* Reading, MA: Addison-Wesley.

Steiner, C. (1974). *Scripts people live: Transactional analysis of life scripts.* New York: Grove Press.

Woollams, S., & Brown, M. (1978). *Transactional analysis.* Dexter, MI: Huron Valley Institute Press.

 ## Psychoanalytic and Psychodynamic counseling

Sigmund Freud established the foundation from which all counseling approaches evolved. His ideas about the unconscious and about personality development led to ingenious counseling techniques and motivated opponents to create such radically different approaches as behavioral counseling and rational/emotive therapy. Much of the work of the psychoanalytic counselor involves making unconscious material conscious, thereby helping counselor and client to gain insight into the mechanisms of psychological adjustment. However, this process is frequently anxiety-provoking to clients who resist self-awareness by means of various defense mechanisms. More than do adherents of any other approach, psychoanalytic counselors emphasize the role of past parent/child transactions and foster the recreation of this relationship in counseling. The projection onto the counselor of clients' feelings toward their parents is called transference. By bringing transference into the open, clients gain new understanding of their psychological processes and ameliorate troubling symptoms.

References

Brill, A. (Ed.). (1938). *The basic writings of Sigmund Freud.* New York: Modern Library.

Chapman, A. H. (1978). *The treatment techniques of Harry Stack Sullivan.* New York: Brunner/Mazel.

Freud, A. (1946). *The ego and the mechanisms of defense.* New York: International Universities Press.

Jung, C. (1959). *Basic writings*. New York: Modern Library.

Kernberg, O. (1981). *Object relations theory and clinical psychoanalysis*. New York: Aronson.

Langs, R. (1973). *The techniques of psychoanalytic psychotherapy* (Vol. 1). New York: Aronson.

Malcolm, J. (1981). *Psychoanalysis, the impossible profession*. New York: Knopf.

Menninger, K., and Holzman, P. (1994). *Theory of psychoanalytic technique*. New York: Basic Books.

Shapiro, D. (1965). *Neurotic styles*. New York: Basic Books.

St. Clair, M. (1987). *Object relations and self-psychology: An introduction*. Pacific Grove, CA: Brooks/Cole.

Strupp, H., & Binder, J. (1985). *Psychotherapy in a new key: A guide to time-limited dynamic psychotherapy*. New York: Basic Books.

Sullivan, H. S. (1953). *The collected works of Harry Stack Sullivan, M. D.* New York: Norton

Wachtel, P. (1977). *Psychoanalysis and behavior therapy: Toward an integration*. New York: Basic Books.

 ## Existential counseling

Existential counselors examine the role of what many consider to be abstract, philosophical issues in the psychological lives of individuals. More than counselors using any other approach, existential counselors eschew technique in favor of grappling with the basic dimensions of life and death.

Thus, people can be considered in terms of *being* (which corresponds to an awareness of oneself) and *nonbeing* (a loss of identity, perhaps caused by conformity). Rollo May, an eminent existential counselor, describes anxiety as the experience of the threat of imminent nonbeing. As do person-centered and Gestalt counselors, existentialists see personal choice and volition as basic facts of human existence. According to existential counselors, clients seek counseling to expand their psychological worlds. As an existential counselor, one's job is to be authentic, to expose oneself to clients so that the clients can become aware of similar qualities in themselves. Along with many behaviorists, existential counselors believe that knowledge and insight follow behavior change, not vice versa.

References

Frankl, V. (1984). *Man's search for meaning.* (3rd ed.) New York: Washington Square Press.

Jourard, S. (1971). *The transparent self* (rev. ed.). New York: Van Nostrand Reinhold.

Keen, E. (1970). *Three faces of being: Toward an existential clinical psychology.* New York: Appleton-Century-Crofts.

May, R. (1983). *The discovery of being.* New York: Norton.

Sartre, J. P. (1971). *Being and nothingness.* New York: Bantam Books.

Watson, R. (1977). An introduction to humanistic psychotherapy. In S. Morse & R. Watson (Eds.), *Psychotherapies: A comparative casebook.* New York: Holt, Rinehart & Winston.

Yalom, I. (1989). *Love's executioner and other tales of psychotherapy.* New York: Basic Books.

 ## Group counseling

Group counselors may subscribe to any of the counseling approaches described previously; that is, you may find Gestalt groups, person-centered groups, and cognitive/behavioral groups. What all group counselors share in common is a recognition of the benefits of working with more than one client at a time.

If clients can be placed in a group on the basis of relatively similar problems, work can be done more efficiently than it can in one-to-one counseling. In most groups, benefits occur from interactions among group members, not from working with the counselor per se. With therapeutic groups, the group counselor is more a facilitator than an active participant or leader. Skilled group counselors supply moderate amounts of group rules and emotional challenges, along with high amounts of support and interpretation of group processes (Lieberman, Yalom, & Miles, 1973).

References

Bednar, R. L., & Kaul, T. J. (1994). Experiential group research: Can the canon fire? In A. E. Bergin & S. L. Garfield (Eds.), *Handbook of psychotherapy and behavior change* (4th ed., pp. 631–663). New York: Wiley.

Fuhriman, A., & Burlingame, G. M. (1994). *Handbook of group psychotherapy.* New York: Wiley.

Gazda, G. (1989). *Group counseling: A developmental approach* (4th ed.). Boston: Allyn & Bacon.

Hansen, J., Warner, R., & Smith, E. (1992). *Group counseling: Theory and process* (5th ed.). Boston: Houghton Mifflin.

Schneider Corey, M., & Corey, G. (1992). *Groups: Process and practice* (4th ed.). Pacific Grove, CA: Brooks/Cole.

Yalom, I. (1995). *The theory and practice of group psychotherapy* (4th ed.). New York: Basic Books.

 ## Family/systems counseling

Family and systems approaches are relatively new approaches to counseling. In contrast to traditional methods, which focus on processes within the individual as the target of change, family and systems counselors chart the influence of social systems on clients. The goal of the family counselor is to detect and understand the methods that family members use to communicate among themselves, maintain the family structure, and help or hinder members' growth. Many systems counselors share with Gestalt counselors an interest in psycholinguistics, the study of how language influences what we think, do, and feel.

References

Bandler, R., & Grinder, J. (1975). *The structure of magic.* Palo Alto, CA: Science and Behavior Books.

Bateson, G. (1972). *Steps to an ecology of mind.* New York: Ballantine.

Bowen, M. (1978). *Family therapy in clinical practice.* New York: Aronson.

deShazer, S. (1982). *Patterns of brief family therapy.* New York: Guilford.

Haley, J. (1987). *Problem-solving therapy* (2nd ed.). San Francisco: Jossey-Bass.

Madanes, C. (1984). *Behind the one-way mirror: Advances in the practice of strategic therapy.* San Francisco: Jossey-Bass.

Mikesell, R. H., Lusterman, D., & McDaniel, S. H. (Eds.) (1995). *Integrating family therapy*. Washington, DC: American Psychological Association.

Minuchin, S., & Fishman, H. C. (1981). *Family therapy techniques*. Cambridge, MA: Harvard University Press.

Napier, A., & Whitaker, C. (1978). *The family crucible*. New York: Harper & Row.

Satir, V., & Baldwin, M. (1983). *Satir step by step*. Palo Alto, CA: Science and Behavior Books.

Watzlawick, P., Weakland, J., & Fisch, R. (1974). *Change: Principles of problem formation and problem resolution*. New York: Norton.

Worden, M. (1994). *Family therapy basics*. Pacific Grove, CA: Brooks/Cole.

 ## Brief therapy

Brief therapy (BT) can be considered one of the newest and most ambiguous approaches to counseling. It has been adopted in recent years largely because of economics—that is, because of the widespread implementation of managed care approaches whose major emphasis often appears to be limiting health care costs. However, BT is ambiguous along several dimensions: (a) proponents of different approaches argue that BT can be comprised of one session or weekly sessions over several years, and (b) BT often appears to be simply a shortened version of other counseling approaches. What does seem common across its different forms is that BT counselors plan sessions more than other counselors, and are more active and directive. Reduction of symptoms (for example, psychological distress) and restoration of functioning are primary goals; clients with clearly defined complaints are most appropriate.

While BT by definition focuses on time limits and efficiency, there is also a recognition that clients return periodically to counseling. Budman (1981) suggests that problems such as depression, stress, and eating disorders "are often chronic in nature and are characterized by their persistent or intermittent nature" (p. 462). Frank (1991) reports that depressed patients, treated in 16 sessions and continued intermittent contact, had fewer relapses that those with only the 16 initial sessions. Thus, Budman (1981) maintains that "time-effective therapies are those that emphasize the efficient, effective use of time, while considering the relative cost, outcome, and benefits of services offered" (p. 463, italics deleted).

Research examining the effects of the number of counseling sessions on outcome has provided no clear consensus. A recent Consumer Reports survey ("Does Therapy Help?, November 1995) of 4000 readers who had previous counseling found that length of therapy was associated with more improvement. But other reviews have suggested no differences or advantages for BT (see Barber, 1994; Steenbarger, 1994). Thus, like most areas in counseling, BT clearly requires more study.

References

Atlas, J. A. Crisis and acute brief therapy with adolescents. *Psychiatric Quarterly, 65,* 79–87.

Barber, J. P. (1994). Efficacy of short-term dynamic psychotherapy: Past, present, and future. *Journal of Psychotherapy Practice & Research, 3,* 108–121.

Budman, S. H. (1981). Brief therapy in the year 2000 and beyond: Looking back while looking forward. In S. H. Budman (Ed.), *Forms of brief therapy* (pp. 461–470). New York: Guilford.

Crits-Christoph, P. (1992). The efficacy of brief dynamic psychotherapy: A meta-analysis. *American Journal of Psychiatry, 149,* 151–158.

Davanloo, H. (1980). *Short-term dynamic psychotherapy.* New York: Jason Aronson.

Does therapy help? (1995, November). *Consumer Reports, 60,* 734–739.

Frank, E. (1991). Interpersonal psychotherapy as a maintenance treatment for patients with recurrent depression. *Psychotherapy, 28,* 259–266.

Malan, D. H. (1976). *The frontier of brief psychotherapy.* New York: Plenum.

McCullough, L., Winston, A., Farber, B., Porter, F., Pollack, J., Laikin, M., Vingiano, W., & Trujillo, M. (1991). The relationship of patient-therapist interaction to outcome in brief psychotherapy. *Psychotherapy, 28,* 525–533.

Murphy, J. J. (1994). Brief therapy for school problems. *School Psychology International, 15,* 115–131.

Steenbarger, B. N. (1994). Duration and outcome in psychotherapy: An integrative review. *Professional Psychology: Research and Practice, 25*, 111–119.

Strupp, H., & Binder, J. (1985). *Psychotherapy in a new key: A guide to time-limited dynamic psychotherapy.* New York: Basic Books.

Research on counseling and psychotherapy

Research in counseling focuses on process and outcome. The basic outcome question is this: Is counseling effective? Most would say yes on the basis of previous research. Perhaps the best result of all the research conducted in counseling and psychotherapy is that we are continually asking better questions (such as, What client characteristics interact with what type of treatment to produce the best outcome?), and we are continually improving the sophistication of the research employed to seek answers to these questions.

Unfortunately, one might argue that improved research sophistication is the most detectable outcome of past decades of counseling research; that is, relatively few counselors pay attention to research beyond their training years, partially because research results often appear to add little to current practice. However, research, theory, and practice must be increasingly integrated if counseling is to progress as a science.

References

Dawes, R. M. (1994). *House of cards: Psychology and psychotherapy built on myth.* New York: Fress Press.

Eysenck, H. (1952). The effects of psychotherapy: An evaluation. *Journal of Consulting Psychology, 16*, 319–324.

Garfield, S., & Bergin, A. (1994). *Handbook of psychotherapy and behavior change: An empirical analysis* (4th ed.). New York: Wiley.

Gelso, C. (1979). Research in counseling: Methodological and professional issues. *The Counseling Psychologist, 8*, 7–35.

Russell, R. L. (1994). *Reassessing psychotherapy research.* New York: Guilford.

Smith, M., & Glass, G. (1977). Meta-analysis of psychotherapy outcome studies. *American Psychologist, 32,* 752–760.

 Other important sources

Not all important publications fit neatly into the categories already listed. The following sources are also worthy of your attention.

References

Adler, A. (1969). *The practice and theory of individual psychology* (2nd ed.). Totowa, NJ: Littlefield.

American Psychiatric Association. (1994). *Diagnostic and statistical manual of mental disorders* (4th ed., Rev.). Washington, DC: Author.

Glasser, W. (1975). *Reality therapy.* New York: Harper & Row.

Lazarus, A. (1981). *The practice of multimodal therapy.* New York: McGraw-Hill.

Lerner, H. G. (1989). *The dance of anger: A woman's guide to changing the patterns of intimate relationships.* New York: Perennial Library.

Zeig, J. (1987). *The evolution of psychotherapy.* New York: Brunner/ Mazel.

References

Abramowitz, S., Weitz, L., Schwartz, J., Amura, S., Gomes, B., & Abramowitz, C. (1975). Comparative counselor inferences toward women with medical school aspirations. *Journal of College Student Personnel, 16,* 128–130.

American Association for Counseling and Development. (1988). *Ethical standards.* Alexandria, VA: Author.

American Psychological Association. (1992). Ethical principles of psychologists. In *Directory of the American Psychological Association.* Washington, DC: Author.

Anton, J., Dunbar, J., & Friedman, L. (1976). Anticipation training in the treatment of depression. In J. Krumboltz & C. Thoresen (Eds.), *Counseling methods.* New York: Holt, Rinehart & Winston.

Bandler, R., & Grinder, J. (1975). *The structure of magic.* Palo Alto, CA: Science and Behavior Books.

Bandura, A. (1977). Self-efficacy theory: Toward a unifying view of behavioral change. *Psychological Review, 84,* 191–215.

Barber, J. P. (1994). Efficacy of short-term dynamic psychotherapy: Past, present, and future. *Journal of Psychotherapy Practice & Research, 3,* 108–121.

Barrett-Lennard, G. (1974). Experiential learning groups. *Psychotherapy: Theory, Research, and Practice, 11,* 71–75.

Bartlett, W., Goodyear, R., & Bradley, F. (Eds.). (1983). Supervision in counseling II. *The Counseling Psychologist, 11*(1).

Bednar, R., & Kaul, T. (1978). Experiential group research: Current perspectives. In S. Garfield & A. Bergin (Eds.), *Handbook of psychotherapy and behavior change: An empirical analysis.* New York: Wiley.

Bergin, A., & Lambert, M. (1978). The evaluation of therapeutic outcomes. In S. Garfield & A. Bergin (Eds.), *Handbook of psychotherapy and behavior change: An empirical analysis.* New York: Wiley.

Bowen, M. (1978). *Family therapy in clinical practice.* New York: Jason Aronson.

Brammer, L., & Shostrom, E. (1989). *Therapeutic psychology* (5th ed.). Englewood Cliffs, NJ: Prentice-Hall.

Brill, A. (1938). *The basic writings of Sigmund Freud.* New York: Modern Library.

Broverman, I., Broverman, D., Clarkson, F., Rosenkrantz, P., & Vogel, S. (1970). Sex-role stereotypes and clinical judgments of mental health. *Journal of Consulting and Clinical Psychology, 34,* 1–7.

Brown, L. (1994). *Subversive dialogues: Theory in feminist therapy.* New York: Basic Books.

Budman, S. H. (1981). Brief therapy in the year 2000 and beyond: Looking back while looking forward. In S. H. Budman (Ed.), *Forms of brief therapy* (pp. 462–470). New York: Guilford.

Butcher, J., & Koss, M. (1978). Research on brief and crisis-oriented psychotherapies. In S. Garfield & A. Bergin (Eds.), *Handbook of psychotherapy and behavior change: An empirical analysis.* New York: Wiley.

Caplan, G. (1961). *An approach to community mental health.* New York: Grune and Stratton.

Carter, R. T. (1991). Cultural values: A review of empirical research and implications for counseling. *Journal of Counseling & Development, 70,* 164–173.

Casas, J. M. (1984). Policy, training, and research in counseling psychology: The racial/ethnic minority perspective. In S. Brown & R. Lent (Eds.), *Handbook of counseling psychology.* New York: Wiley.

Christensen, A., & Jacobson, N. S. (1994). Who (or what) can do psychotherapy: The status and challenge of nonprofessional therapies. *Psychological Science, 5,* 8–14.

Clunis, D. M., & Green, G. D. (1988). *Lesbian couples.* Seattle: Seal Press.

Corey, G., & Schneider Corey, M. (1993). *Issues and ethics in the helping professions* (4th ed.). Pacific Grove, CA: Brooks/Cole.

Cormican, J. (1978, March). Linguistic issues in interviewing. *Social Casework, 145–151.*

Cormier, W., & Cormier, L. (1991). *Interviewing strategies for helpers: Fundamental skills and cognitive behavioral interventions* (3rd ed.). Pacific Grove, CA: Brooks/Cole.

Crits-Christoph, P. (1992). The efficacy of brief dynamic psychotherapy: A meta-analysis. *American Journal of Psychiatry, 149,* 151–158.

deShazer, S. (1982). *Patterns of brief family therapy.* New York: Guilford.

Does therapy help? (1995, November). *Consumer Reports, 60,* 734–739.

Egan, G. (1990). *The skilled helper: A systematic approach to effective helping* (4th ed.). Pacific Grove, CA: Brooks/Cole.

Egan, G. (1994). *The skilled helper: A problem-management approach to helping* (5th ed.). Pacific, Grove, CA: Brooks/Cole.

Eisler, R. (1976). Assertive training in the work situation. In J. Krumboltz & C. Thoresen (Eds.), *Counseling methods.* New York: Holt, Rinehart & Winston.

Ellis, A., & Grieger, R. (1977). *Handbook of rational-emotive therapy.* New York: Springer.

Ellis, A., & Harper, R. (1976). *A new guide to rational living.* North Hollywood, CA: Wilshire Book Company.

Flaskerud, J. H. (1990). Matching client and therapist ethnicity, language, and gender: A review of research. *Issues in Mental Health Nursing, 11,* 321–336.

Forsyth, D., & Strong, S. (1986). The scientific study of counseling and psychotherapy: A unificationist view. *American Psychologist, 41,* 113–119.

Frank, E. (1991). Interpersonal psychotherapy as a maintenance treatment for patients with recurrent depression. *Psychotherapy, 28,* 259–266.

Frank, J. (1971). Therapeutic factors in psychotherapy. *American Journal of Psychotherapy, 25,* 350–361.

Freudenberger, H. (1974). Staff burnout. *Journal of Social Issues, 30,* 159–165.

Gelso, C. (1979). Research in counseling: Methodological and professional issues. *The Counseling Psychologist, 8,* 7–35.

Glidden-Tracey, C. E., & Wagner, L. (1995). Gender salient attribute* treatment interaction effects on ratings of two anlogue counselors. *Journal of Counseling Psychology, 42,* 223–231.

Goldfried, M. R. (1983). The behavior therapist in clinical practice. *Behavior Therapist, 6,* 45–46.

Goldman, L. (1976). A revolution in counseling research. *Journal of Counseling Psychology, 23,* 543–552.

Greenson, R. (1965). The working alliance and the transference neurosis. *Psychoanalytic Quarterly, 34,* 155–181.

Gunzburger, D., Henggeler, S., & Watson, S. (1985). Factors related to premature termination of counseling relationships. *Journal of College Student Personnel, 26,* 456–460.

Gurman, A., & Kniskern, D. (1978). Research on marital and family therapy: Progress, perspective, and prospect. In S. Garfield & A. Bergin (Eds.), *Handbook of psychotherapy and behavior change: An empirical analysis.* New York: Wiley.

Hansen, J., Rossberg, R. R. & Cramer, S. H. (1994). *Counseling: Theory and process,* (4th ed.). Boston: Allyn & Bacon.

Hartman, D. P. (1984). Assessment strategies. In D. H. Barlow & M. Hersen (Eds.), *Single case experimental designs* (pp. 107–129). New York: Pergamon.

Helms, J. (1979). Perceptions of a sex-fair counselor and her client. *Journal of Counseling Psychology, 26,* 504–513.

Heppner, P. P., Kivlighan, D. M., Jr., & Wampold, B. E. (1992). *Research design in counseling.* Pacific Grove, CA: Brooks/Cole.

Herr, E., & Best, P. (1984). Computer technology and counseling: The role of the profession. *Journal of Counseling and Development, 63,* 192–195.

Hodgson, R., & Rachman, S. (1974). II. Desynchrony in measures of fear. *Behaviour Research and Therapy, 12,* 319–326.

Hoehn-Saric, R., Frank, J., Imber, S., Nash, E., Stone, A., & Battle, C. (1964). Systematic preparation of patients for psychotherapy. I. Effects on therapy behavior and outcome. *Journal of Psychiatric Research, 2,* 267–281.

Ivey, A. (1980). Counseling 2000: Time to take charge! *The Counseling Psychologist, 8,* 12–16.

Jourard, S. (1971). *Self-disclosure: An experimental analysis of the transparent self.* New York: Wiley.

Kahn, M. (1990). *Between therapist and client.* New York: W. H. Freeman.

Karasu, T. (1986). The specificity versus nonspecificity dilemma: Toward identifying therapeutic change agents. *American Journal of Psychiatry, 143,* 687–695.

Kiesler, D. J. (1971). Experimental designs in psychotherapy research. In A. Bergin & S. Garfield (Eds.), *Handbook of psychotherapy and behavior change.* New York: Wiley.

Kilburg, R., Nathan, P., & Thoreson, R. (1986). *Professionals in distress.* Washington, DC: American Psychological Association.

Krumboltz, J. (Ed.). (1966). *Revolution in counseling.* Boston: Houghton Mifflin.

Kuhn, T. (1970). *The structure of scientific revolution.* Chicago: University of Chicago Press.

Landman, J., & Dawes, R. (1982). Psychotherapy outcome. *American Psychologist, 37,* 504–516.

Langs, R. (1973). *The techniques of psychoanalytic psychotherapy* (Vol. 1). New York: Aronson.

Lieberman, M., Yalom, I., & Miles, M. (1973). *Encounter groups: First facts.* New York: Basic Books.

MacKinnon, R., & Michels, R. (1971). *The psychiatric interview in clinical practice.* Philadelphia: Saunders.

Mahoney, M. (1987, August 31). *Plasticity and power: Emerging emphases in theories of human change.* Paper presented at the 95th Annual Meeting of the American Psychological Assoc., New York.

Mayerson, N. (1984). Preparing clients for group therapy: A critical review and theoretical formulation. *Clinical Psychology Review, 4,* 191–213.

McCarthy, P. (1982). Differential effects of counselor self-referent responses and counselor status. *Journal of Counseling Psychology, 29,* 125–131.

McCarthy, P., & Betz, N. (1978). Differential effects of self-disclosing versus self-involving counselor statements. *Journal of Counseling Psychology, 25,* 251–256.

Meehl, P. (1956). Wanted—A good cookbook. *American Psychologist, 11,* 262–272.

Meehl, P. (1973). Why I no longer attend clinical case conferences. In *Psychodiagnosis: Selected papers.* Minneapolis: University of Minnesota Press.

Meier, S. (1983). Toward a theory of burnout. *Human Relations, 36,* 899–910.

Meier, S. (1987). An unconnected special issue. *American Psychologist, 42,* 881.

Mikesell, R. H., Lusterman, D., & McDaniel, S. H. (Eds.) (1995). *Integrating family therapy.* Washington, DC: American Psychological Association.

Minuchin, S. (1974). *Families and family therapy.* Cambridge, MA: Harvard University Press.

Minuchin, S., & Fishman, H. C. (1981). *Family therapy techniques.* Cambridge, MA: Harvard University Press.

Mohr, D. C. (1995). Negative outcome in psychotherapy: A critical review. *Clinical Psychology: Science & Practice, 2,* 1–27.

Moxley, D. (1989). *The practice of case management.* Newbury Park, CA: Sage Publications.

Orlinsky, D., & Howard, K. (1978). The relation of process to outcome in psychotherapy. In S. Garfield & A. Bergin (Eds.), *Handbook of psychotherapy and behavior change: An empirical analysis.* New York: Wiley.

Paul, G. L. (Ed.). (1986). *Assessment in residential treatment settings.* Champaign, IL: Research Press.

Pope, B. (1979). *The mental health interview.* New York: Pergamon Press.

Principles concerning the counseling and therapy of women. (1978). *The Counseling Psychologist, 7,* 4.

Puryear, D. (1979). *Helping people in crisis: A practical family-oriented approach to effective crisis intervention.* San Francisco: Jossey-Bass.

Rachman, S., & Hodgson, R. (1974). Synchrony and desynchrony in fear and avoidance. *Behaviour Research and Therapy, 12,* 311–318.

Richardson, M., & Johnson, M. (1984). Counseling women. In S. Brown & R. Lent (Eds.), *Handbook of counseling psychology.* New York: Wiley.

Robertiello, R., & Schoenewolf, G. (1987). *101 Common therapeutic blunders.* Northvale, NJ: Aronson.

Rogers, C. (1963). Psychotherapy today, or where do we go from here? *American Journal of Psychotherapy, 17,* 5–16.

Rosenthal, D., & Frank, J. (1958). Psychotherapy and the placebo effect. In C. Reed, I. Alexander, & S. Tomkins (Eds.), *Psychopathology: A source book.* Cambridge, MA: Harvard University Press.

Russell, R. (1970). Black perceptions of guidance. *Personnel & Guidance Journal, 48,* 721–728.

Satir, V. (1988). *New people-making* (2nd ed.). Palo Alto, CA: Science and Behavior Books.

Schmidt, L., & Meara, N. (1984). Ethical, professional, and legal issues in counseling psychology. In S. Brown & R. Lent (Eds.), *Handbook of counseling psychology.* New York: Wiley.

Schneider Corey, M., & Corey, G. (1992). *Groups: Process and practice* (4th ed.). Pacific Grove, CA: Brooks/Cole.

Shertzer, B., & Stone, S. (1980). *Fundamentals of counseling* (3rd ed.). Boston: Houghton Mifflin.

Slaikeu, K. (1990). *Crisis intervention* (2nd ed.). Boston: Allyn & Bacon.

Smith, M., & Glass, G. (1977). Meta-analysis of psychotherapy outcome studies. *American Psychologist, 32,* 752–760.

Sommers-Flanagan, J., & Sommers-Flanagan, R. (1993). *Foundations of therapeutic interviewing*. Boston: Allyn & Bacon.

Staats, A. (1983). *Psychology's crisis of disunity: Philosophy and method for a unified science*. New York: Praeger.

Stanard, R., & Hazler, R. (1995). Legal and ethical implications of HIV and duty to warn for counselors: When does Tarasoff apply? *Journal of Counseling and Development, 73,* 397–400.

Steenbarger, B. N. (1994). Duration and outcome in psychotherapy: An integrative review. *Professional Psychology: Research and Practice, 25,* 111–119.

Strunk, W., Jr., & White, E. B. (1979). *The elements of style* (3rd ed.). New York: Macmillan.

Stuart, R. (1974). Teaching facts about drugs: Pushing or preventing? *Journal of Educational Psychology, 66,* 189–201.

Sue, D. W. (1990a). *Counseling the culturally different: Theory and practice*. New York: Wiley.

Sue, D. W. (1990b). Culture-specific strategies in counseling: A conceptual framework. *Professional Psychology: Research & Practice, 21,* 424–433.

Sullivan, H. S. (1970). *The psychiatric interview*. New York: Norton.

Teyber, E. (1992). *Interpersonal process in psychotherapy: A guide for clinical training* (2nd ed.). Pacific Grove, Ca: Brooks/Cole.

Van Kaam, A. (1966). *The art of existential counseling*. Wilkes-Barre, PA: Dimension Books.

Varenhorst, B. (1984). Peer counseling: Past promises, current status, and future directions. In S. Brown & R. Lent (Eds.), *Handbook of counseling psychology*. New York: Wiley.

Watzlawick, P., Weakland, J., & Fisch, R. (1974). *Change: Principles of problem formation and problem resolution*. New York: Norton.

Whiteley, J. (Ed.). (1982). Supervision in counseling I. *The Counseling Psychologist, 10,* 1.

Wolpe, J. (1990). *The practice of behavior therapy* (4th ed.). New York: Pergamon Press.

Worden, M. (1994). *Family therapy basics*. Pacific Grove, CA: Brooks/Cole.

Worell, J. (1992). *Feminist perspectives in therapy*. New York: Wiley.

Yalom, I. (1995). *The theory and practice of group psychotherapy* (4th ed.). New York: Basic Books.

Zetzel, E. (1956). Current concepts of transference. *International Journal of Psychoanalysis, 37,* 369–376.

Index

TO THE OWNER OF THIS BOOK:

We hope that you have found *Elements of Counseling,* Third Edition, useful. So that this book can be improved in a future edition, would you take the time to complete this sheet and return it? Thank you.

School and address: _____

Department: _____

Instructor's name: _____

1. What I like most about this book is: _____

_____

_____

2. What I like least about this book is: _____

_____

_____

3. My general reaction to this book is: _____

_____

_____

4. The name of the course in which I used this book is: _____

_____

5. In the space below, or on a separate sheet of paper, please write specific suggestions for improving this book and anything else you'd care to share about your experience in using the book.

_____

_____

_____

_____

Optional:

Your name: _____ Date: _____

May Brooks/Cole quote you, either in promotion for *Elements of
Counseling,* Third Edition, or in future publishing ventures?

Yes: _____ No: _____

Sincerely,

Scott T. Meier
Susan R. Davis